D0602931

ALGEBRA SUCCESS

IN 20 MINUTES A DAY

ALGEBRA SUCCESS
IN 20 MINUTES A DAY

Second Edition

NEW YORK

Published in the United States by LearningExpress, LLC, New York.

Library of Congress Cataloging-in-Publication Data:
Algebra success in 20 minutes a day.—2nd ed.
p. cm.
Rev. ed. of: Algebra success in 20 minutes a day / Barbara Jund. 1st ed. © 2000
ISBN 1-57685-486-8
1. Algebra—Study and teaching. I. Jund, Barbara. Algebra success in 20 minutes a day.
II. Title: Algebra success in twenty minutes a day.
QA159.J59 2005
512' .007—dc22

2005040829

Printed in the United States of America

9 8 7 6 5 4 3 2 1

Second Edition

For information on LearningExpress, other LearningExpress products, or bulk sales, please write us at:
LearningExpress
55 Broadway
8th Floor
New York, NY 10006

Or visit us at:
www.learnatest.com

Contents

Introduction

If you have never taken an algebra course and now find that you need to know algebra, this is the book for you. If you have already taken an algebra course but felt like you never understood what the teacher was trying to tell you, this book can teach you what you need to know. If it has been awhile since you have taken an algebra course and you need to refresh your skills, this book will review the basics and reteach you the skills you may have forgotten. Whatever your reason for needing to know algebra, *Algebra Success* will teach you what you need to know. It gives you the basics of an Algebra I course in clear and straightforward lessons that you can do at your own pace.

Many math teachers often hear the comment, "I was never very good in math." If you didn't take algebra because you thought it was too hard, you will be surprised to find out how easy it is. If you took algebra but didn't understand it, when you finish this book, you won't believe how easy algebra can be.

Algebra is math with *variables,* numbers whose actual value is not yet known. The ability to calculate with the unknown makes algebra essential for science, business, and all the technologies of the future that are still being worked out. If all you can do is arithmetic, you are limited to the ever-dwindling pool of jobs that are slowly being replaced by those technologies.

▶ Overcoming Math Anxiety

Do you like math or do you find math an unpleasant experience? It is human nature for people to like what they are good at. Generally, people who dislike math have not had much success with math.

If you have struggled with math, ask yourself why. Was it because the class went too fast? Did you have a chance to fully understand a concept before you went on to a new one? One of the comments students frequently make is, "I was just starting to understand, and then the teacher went on to something new." That is why *Algebra Success* is self-paced. You work at your own pace. You go on to a new concept only when you are ready.

Algebra Success goes straight to the basics using common, everyday language. Great care was taken to explain concepts in clear language so that you would not get lost in mathematical jargon. Only the algebra terms that you need to function in a basic algebra course were included.

When you study the lessons in this book, the only person you have to answer to is "you." You don't have to pretend you know something when you don't truly understand. You get to take the time you need to understand everything before you go on to the next lesson. You have truly learned something only if you thoroughly understand it. Merely completing a lesson does not mean you understand it. When you go through a lesson, work for understanding. Take as much time as you need to understand the examples. Check your work with the answers as you progress through the lesson. If you get the right answer, you are on the right track! If you finish a lesson and you don't feel confident that you fully understand the lesson, do it again. Athletes and musicians practice a skill until they perfect it. Repetition works for mathematicians, too. Remember the adage, "Practice makes perfect." You might think you don't want to take the time to go back over something again. However, making sure you understand a lesson completely may save you time in future lessons. Rework problems you missed to make sure you don't make the same mistakes again.

► How to Use This Book

Algebra Success teaches basic algebra concepts in 20 self-paced lessons. The book also includes a pretest, a posttest, a glossary of mathematical terms, and an appendix of additional resources for further study. Before you begin Lesson 1, take the pretest. The pretest will assess your current algebra abilities. You'll find the answer key for the pretest at the end of the book. Each answer includes the lesson number that the problem is testing. This will be helpful in determining your strengths and weaknesses. After taking the pretest, move on to Lesson 1.

Each lesson offers detailed explanations of a new concept. There are numerous examples with step-by-step solutions. As you proceed through a lesson, you will find tips and shortcuts that will help you learn a concept. Each new concept is followed by a practice set of problems. The practice problems allow you to practice each new concept without tedious calculations. You will find that most calculations can be done without the use of a calculator. The emphasis is on algebra concepts—not calculations. The answers to the practice problems are in an answer key located at the end of the book. Some lessons include word problems that will illustrate real-life applications of the algebra concept that was studied in the lesson. Algebra is a tool that is used to solve many real-life problems. At the end of each lesson is an exercise called "Skill Building until Next Time." This exercise applies the lesson's topic to an activity you may encounter in your daily life.

As you work through the practice problems in this book, remember that it is extremely important to write out your steps as you work through a problem. When you write out your steps, you are developing your thinking in an organized manner. When you have steps written down on paper, you can see where you made a mistake when a problem was worked incorrectly. If you don't write the steps down on paper, you can only guess where you made the mistake. Good organization develops good math skills!

When you have completed all 20 lessons, take the posttest at the end of the book. The posttest has the same format as the pretest, but the questions are different. Compare the results of the posttest with the results of the pretest you took before you began Lesson 1. What are your strengths? Do you have weak areas? Do you need to spend more time on some concepts, or are you ready to go to the next level?

▶ Make a Commitment

Success does not come without effort. Make the commitment to improve your math skills. Work for understanding. *Why* you do a math operation is as important as *how* you do it. If you truly want to be successful, make a commitment to spend the time you need to do a good job. You can do it! When you achieve algebra success, you have laid the foundation for future challenges and success.

So sharpen that pencil and get ready to begin the pretest!

ALGEBRA SUCCESS

IN 20 MINUTES A DAY

Pretest

Before you begin Lesson 1, you may want to get an idea of what you know and what you need to learn. The pretest will answer some of these questions for you. The pretest consists of 50 multiple-choice questions that cover the topics in this book. While 50 questions can't cover every concept, skill, or shortcut taught in this book, your performance on the pretest will give you a good indication of your strengths and weaknesses. Keep in mind the pretest does not test all the skills taught in this book, but it will tell you the degree of effort you will need to put forth to accomplish your goal of learning algebra.

If you score high on the pretest, you have a good foundation and should be able to work your way through the book quickly. If you score low on the pretest, don't despair. This book will take you through the algebra concepts, step by step. If you get a low score, you may need to take more than 20 minutes a day to work through a lesson. However, this is a self-paced program, so you can spend as much time on a lesson as you need. You decide when you fully comprehend the lesson and are ready to go on to the next one.

Take as much time as you need to do the pretest. When you are finished, check your answers with the answer key at the end of the book. Along with each answer is a number that tells you which lesson of this book teaches you about the algebra skills needed for that question. You will find the level of difficulty increases as you work your way through the pretest.

1.	ⓐ	ⓑ	ⓒ	ⓓ
2.	ⓐ	ⓑ	ⓒ	ⓓ
3.	ⓐ	ⓑ	ⓒ	ⓓ
4.	ⓐ	ⓑ	ⓒ	ⓓ
5.	ⓐ	ⓑ	ⓒ	ⓓ
6.	ⓐ	ⓑ	ⓒ	ⓓ
7.	ⓐ	ⓑ	ⓒ	ⓓ
8.	ⓐ	ⓑ	ⓒ	ⓓ
9.	ⓐ	ⓑ	ⓒ	ⓓ
10.	ⓐ	ⓑ	ⓒ	ⓓ
11.	ⓐ	ⓑ	ⓒ	ⓓ
12.	ⓐ	ⓑ	ⓒ	ⓓ
13.	ⓐ	ⓑ	ⓒ	ⓓ
14.	ⓐ	ⓑ	ⓒ	ⓓ
15.	ⓐ	ⓑ	ⓒ	ⓓ
16.	ⓐ	ⓑ	ⓒ	ⓓ
17.	ⓐ	ⓑ	ⓒ	ⓓ
18.	ⓐ	ⓑ	ⓒ	ⓓ
19.	ⓐ	ⓑ	ⓒ	ⓓ
20.	ⓐ	ⓑ	ⓒ	ⓓ
21.	ⓐ	ⓑ	ⓒ	ⓓ
22.	ⓐ	ⓑ	ⓒ	ⓓ
23.	ⓐ	ⓑ	ⓒ	ⓓ
24.	ⓐ	ⓑ	ⓒ	ⓓ
25.	ⓐ	ⓑ	ⓒ	ⓓ
26.	ⓐ	ⓑ	ⓒ	ⓓ
27.	ⓐ	ⓑ	ⓒ	ⓓ
28.	ⓐ	ⓑ	ⓒ	ⓓ
29.	ⓐ	ⓑ	ⓒ	ⓓ
30.	ⓐ	ⓑ	ⓒ	ⓓ
31.	ⓐ	ⓑ	ⓒ	ⓓ
32.	ⓐ	ⓑ	ⓒ	ⓓ
33.	ⓐ	ⓑ	ⓒ	ⓓ
34.	ⓐ	ⓑ	ⓒ	ⓓ
35.	ⓐ	ⓑ	ⓒ	ⓓ
36.	ⓐ	ⓑ	ⓒ	ⓓ
37.	ⓐ	ⓑ	ⓒ	ⓓ
38.	ⓐ	ⓑ	ⓒ	ⓓ
39.	ⓐ	ⓑ	ⓒ	ⓓ
40.	ⓐ	ⓑ	ⓒ	ⓓ
41.	ⓐ	ⓑ	ⓒ	ⓓ
42.	ⓐ	ⓑ	ⓒ	ⓓ
43.	ⓐ	ⓑ	ⓒ	ⓓ
44.	ⓐ	ⓑ	ⓒ	ⓓ
45.	ⓐ	ⓑ	ⓒ	ⓓ
46.	ⓐ	ⓑ	ⓒ	ⓓ
47.	ⓐ	ⓑ	ⓒ	ⓓ
48.	ⓐ	ⓑ	ⓒ	ⓓ
49.	ⓐ	ⓑ	ⓒ	ⓓ
50.	ⓐ	ⓑ	ⓒ	ⓓ

1. Simplify the expression: $5 - {}^{-}7$.

a. -2
b. 12
c. -12
d. 2

2. Simplify the expression: $2 \cdot -3 \cdot -4$.

a. -24
b. 14
c. 10
d. 24

3. Simplify the expression: $-8 - 11 + 2$.

a. -21
b. -17
c. 5
d. 17

4. Simplify the expression: $\frac{-35}{-7}$.

a. -5
b. 6
c. -6
d. 5

5. Simplify the expression: $2 + {}^{-}3 \cdot 4 - 4 \div 2$.

a. -12
b. 0
c. 2
d. -6

6. Simplify the expression: $6 - 3(-1 + 4 \cdot 3)$.

a. 33
b. -27
c. -21
d. 27

7. Evaluate $3a - bc^2$ when $a = 5$, $b = 2$, and $c = -3$.

a. -3
b. 117
c. 33
d. -21

8. Simplify: $6xy + 8x^2y - 3xy$.

a. $11x^3y^2$
b. $11x^2y$
c. $3xy + 8x^2y$
d. $9xy + 8x^2y$

9. Simplify: $4 - (x - 3) + 8x$.

a. $9x + 7$
b. $-7x + 7$
c. $7x + 7$
d. $1 + 7x$

10. Solve the equation: $x - 8 = -11$.

a. -3
b. 3
c. -19
d. 19

11. Solve the equation: $x - {}^{-}11 = 9$.

a. -2
b. 20
c. -20
d. 2

12. Solve the equation: $-3x = -9$.

a. -3
b. 3
c. 27
d. -27

13. Solve the equation: $\frac{2}{3}x = 6$.

a. 4
b. 2
c. 9
d. 3

14. Solve the equation: $-2x - 1 = 3$.

a. 2
b. -2
c. 1
d. -1

15. Solve the equation: $3x + 6 = -15$.

a. 3
b. –3
c. –11
d. –7

16. Solve the equation: $\frac{x}{4} + 8 = 4$.

a. –48
b. –16
c. –3
d. 16

17. Solve the equation: $3d = 5d - 20$.

a. $-\frac{5}{2}$
b. –10
c. 10
d. $\frac{5}{2}$

18. Solve the equation: $4c - 2 = 8c + 14$.

a. –1
b. –4
c. –3
d. 1

19. Find the area of a trapezoid if $b_1 = 10$ ft., $b_2 = 16$ ft., and the height is 8 ft. Use the formula $A = \frac{1}{2}h(b_1 + b_2)$.

a. 104
b. 168
c. 52
d. 208

20. What amount of money would you have to invest to earn \$2,500 in 10 years if the interest rate is 5%? Use the formula $I = prt$.

a. \$1,250
b. \$50,000
c. \$500
d. \$5,000

21. What is the slope in the equation $y = \frac{2}{3}x + 5$?

a. $\frac{3}{2}$
b. $\frac{2}{3}$
c. 2
d. 5

22. Transform the equation $3x + y = 5$ into slope-intercept form.

a. $y = 3x + 5$
b. $y = -3x + 5$
c. $x = \frac{1}{3}y + 5$
d. $x = -\frac{1}{3}y + 5$

23. Choose the equation that fits the graph.

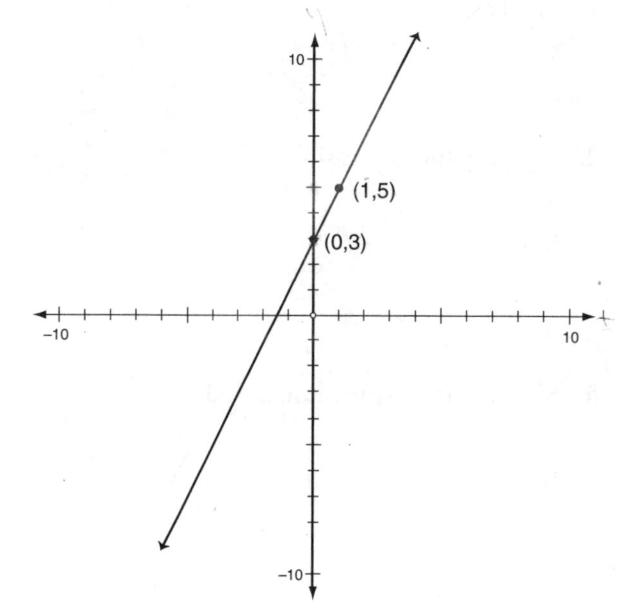

a. $y = 2x + 3$
b. $3x + y = 2$
c. $-3x + y = 2$
d. $y = -2x + 3$

24. Solve the inequality: $4x + 4 > 24$.

a. $x > 7$
b. $x > 5$
c. $x < 5$
d. $x < 7$

25. Solve the inequality: $x + 5 \geq 3x + 9$.

a. $x \geq \frac{7}{2}$
b. $x \geq -2$
c. $x \leq -2$
d. $x \leq 2$

26. Match the graph with the inequality: $y > 4$.

a.

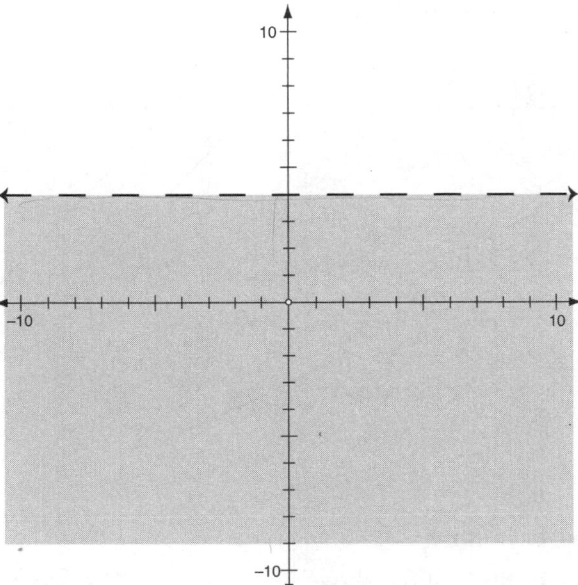

b.

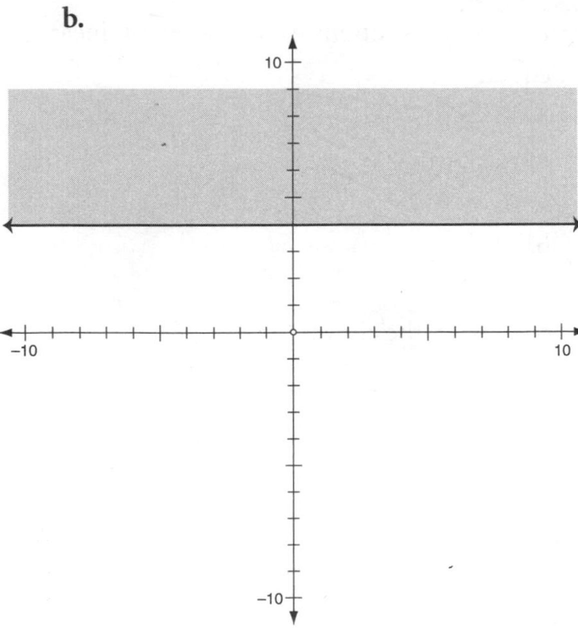

c.

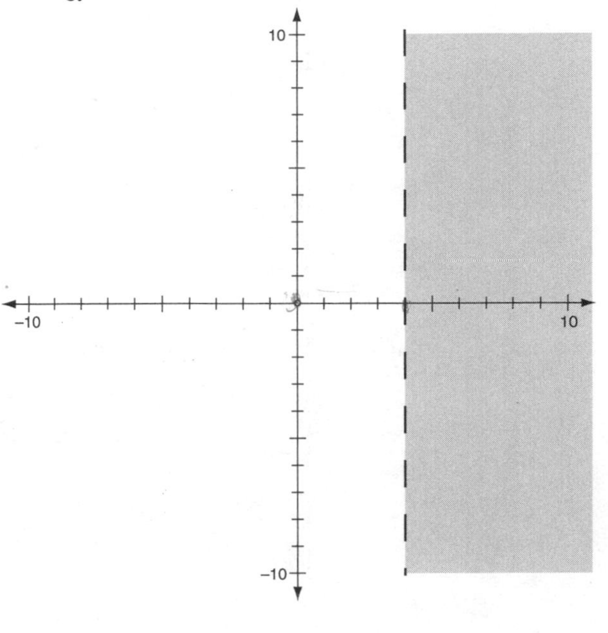

d.

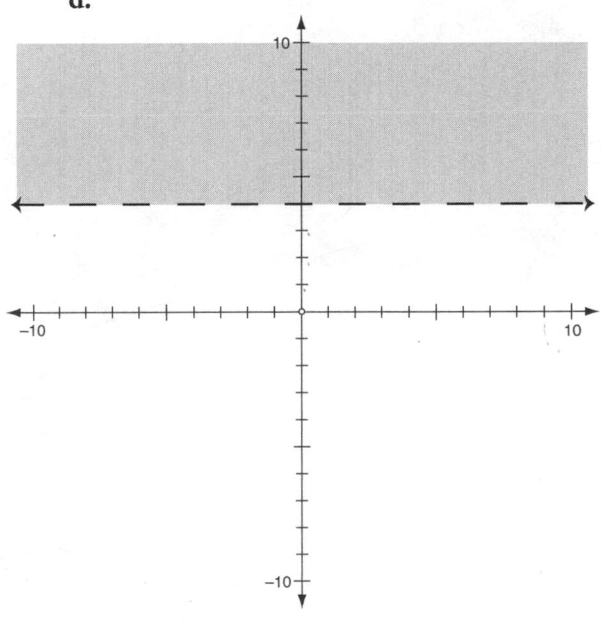

27. Match the inequality with the graph.

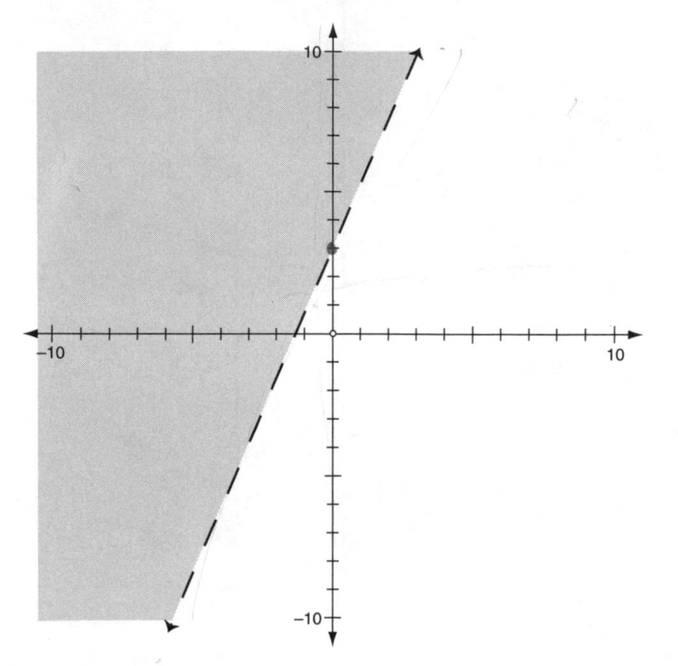

a. $y < 2x + 3$
b. $y \leq 2x + 3$
c. $y > 2x + 3$
d. $y \geq 2x + 3$

28. Determine the number of solutions the system of equations has by looking at the graph.

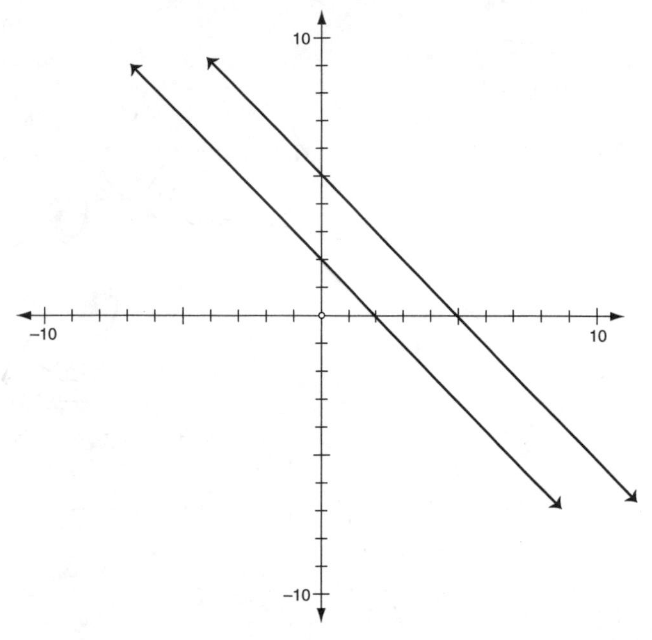

a. 1
b. 0
c. infinite
d. none of the above

29. Use the slope and intercept to determine the number of solutions to the system of linear equations:

$3y + 6 = 2x$

$3y = 2x + 6$

a. 0
b. 1
c. ∞
d. none of the above

30. Select the graph for the system of inequalities:

$y > 2$

$y \leq 2x + 1$

a.

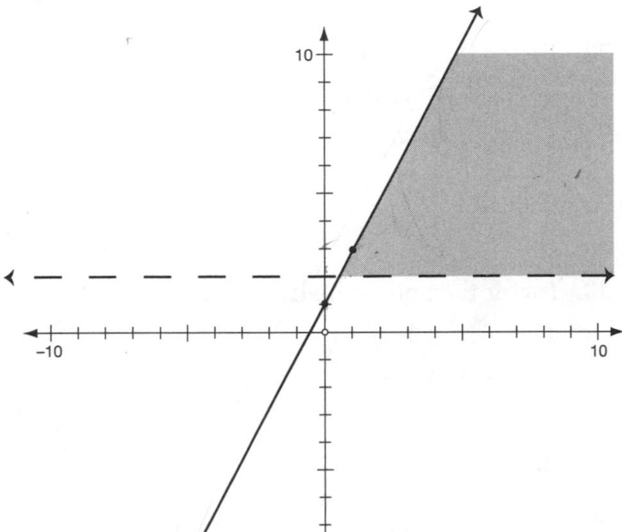

b.

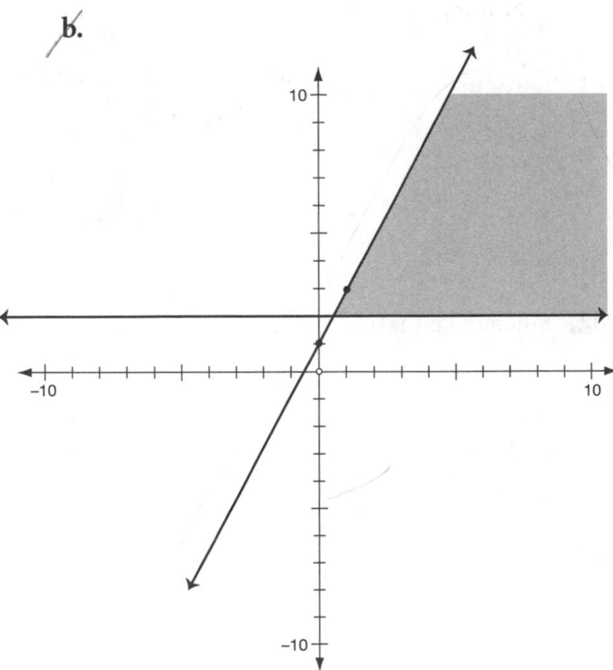

c.

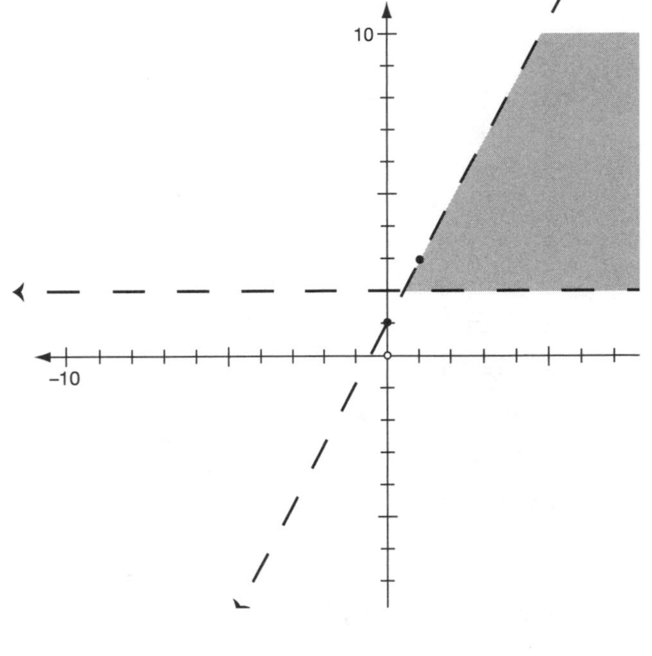

d.

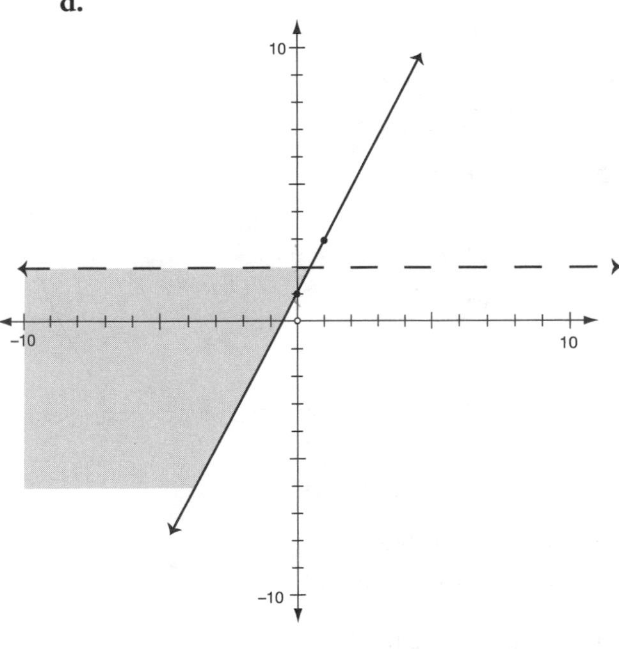

31. Solve the system of equations algebraically:
$2x - y = 10$
$3x + y = 15$
a. (0,5)
b. (5,0)
c. (–5,0)
d. (0,–5)

32. Solve the system of equations algebraically:
$4x - 3y = 10$
$5x + 2y = 1$
a. (4, –3)
b. (1, –2)
c. $(-1, -\frac{1}{3})$
d. $(2, -\frac{2}{3})$

33. Simplify: $2x^2y(3x^3y^2)$.
a. $6x^6y^2$
b. $6x^5y^2$
c. $6x^5y^3$
d. $6x^6y^3$

34. Simplify: $5(2xy^3)^3$.
a. $10x^3y^6$
b. $10x^3y^9$
c. $11x^4y^6$
d. $40x^3y^9$

35. Multiply the polynomials: $2x^2(3x + 4xy - 2xy^3)$.
a. $6x^3 + 8x^2y - 4x^3y^3$
b. $6x^3 + 8x^3y - 4x^3y^3$
c. $6x^3 + 8x^3y - 4x^2y^3$
d. $6x^2 + 8x^2y - 4x^3y^3$

36. Multiply the binomials: $(2x + 3)(x - 2)$.
a. $2x^2 + x - 6$
b. $2x^2 - x + 6$
c. $2x^2 - x - 6$
d. $2x^2 + x + 6$

37. Factor the polynomial: $3a^2b + 6a^3b^2 - 15a^2b^4$.
a. $3a^2b(2ab - 5b^3)$
b. $3a^2b(1 + 2a^2b - 5b^3)$
c. $3a^2b(1 + 2ab + 5b^3)$
d. $3a^2b(1 + 2ab - 5b^3)$

38. Factor the polynomial: $49w^2 - 81$.
a. $(7w + 9)(7w - 9)$
b. $(7w - 9)(7w - 9)$
c. $(7w + 9)(7w + 9)$
d. $(7w - 9)^2$.

39. Factor the polynomial: $x^2 + 3x - 18$.
a. $(x - 2)(x + 9)$
b. $(x + 3)(x - 6)$
c. $(x + 2)(x - 9)$
d. $(x - 3)(x + 6)$

40. Factor the polynomial: $10x^2 + 13x - 3$.
a. $(2x + 3)(5x - 1)$
b. $(2x - 3)(5x + 1)$
c. $(2x + 1)(5x - 3)$
d. $(2x - 1)(5x + 3)$

41. Solve the equation: $3x^2 - 27 = 0$.
a. 0, 3
b. 3, 3
c. 3, –3
d. –3, –3

42. Solve the equation: $2x^2 + x = 3$.
a. $-1, \frac{3}{2}$
b. $-\frac{3}{2}, 1$
c. $3, -\frac{1}{2}$
d. $-3, \frac{1}{2}$

43. Simplify: $-3\sqrt{x} + 2\sqrt{x} + 3\sqrt{y}$.
a. $\sqrt{x} + 3\sqrt{y}$
b. $2\sqrt{xy}$
c. $-\sqrt{x} + 3\sqrt{y}$
d. $-5\sqrt{x} + 3\sqrt{y}$

44. Simplify: $3\sqrt{10xy} \cdot 4\sqrt{6x}$

a. $720x\sqrt{y}$

b. $12\sqrt{60xy}$

c. $24x\sqrt{15y}$

d. $48\sqrt{2xy}$

45. Simplify: $\sqrt{18} + 5\sqrt{2}$.

a. $8\sqrt{20}$

b. $10\sqrt{5}$

c. $8\sqrt{2}$

d. $15\sqrt{2}$

46. Simplify: $\frac{\sqrt{40}}{\sqrt{15}}$

a. $\frac{8}{3}$

b. $\frac{2}{3}\sqrt{2}$

c. $\frac{2}{3}\sqrt{6}$

d. $2\sqrt{10}$

47. Solve the equation: $\sqrt{x} + 3 = 5$.

a. 2

b. 8

c. 4

d. 64

48. Solve the equation: $3\sqrt{x+1} = 15$.

a. 4

b. 24

c. 26

d. 6

49. Use the quadratic formula to solve: $3x^2 + x - 2$.

a. $3, -1$

b. $\frac{2}{3}, -1$

c. $-\frac{2}{3}, 1$

d. $-\frac{1}{6}, \frac{5}{6}$

50. Use the quadratic formula to solve: $4x^2 - 3x - 2$.

a. $\frac{-3 \pm \sqrt{41}}{8}$

b. $\frac{3 \pm \sqrt{41}}{8}$

c. $\frac{-3 \pm \sqrt{17}}{8}$

d. $\frac{3 \pm \sqrt{17}}{8}$

LESSON

Working with Integers

LESSON SUMMARY

Algebra is the branch of mathematics that denotes quantities with letters and uses negative numbers as well as ordinary numbers. In this lesson, you will be working with a set of numbers called *integers*. You use integers in your daily life. A profit is represented with a positive number and a loss is shown using a negative number. This lesson defines integers and explains the rules for adding, subtracting, multiplying, and dividing integers.

▶ What Is an Integer?

Integers are all the positive whole numbers (whole numbers do not include fractions), their opposites, and zero. For example, the opposite of 2 (positive 2) is the number –2 (negative 2). The opposite of 5 (positive 5) is –5 (negative 5). The opposite of 0 is 0. Integers are often called "signed numbers" because we use the positive and negative signs to represent the numbers. The numbers greater than zero are positive numbers, and the numbers less than zero are negative numbers. If the temperature outside is 70°, the temperature is represented with a positive number. However, if the temperature outside is 3 below zero, we represent this number as –3, which is a negative number.

Integers can be represented in this way:

$$\ldots -3, -2, -1, 0, 1, 2, 3, \ldots$$

The three dots that you see at the beginning and the end of the numbers mean the numbers go on forever in both directions. Notice that the numbers get increasingly smaller when you advance in the negative direction and increasingly larger when you advance in the positive direction. For example, –10 is less than –2. The mathematical symbol for less than is “$<$” so we say that $-10 < -2$. The mathematical symbol for greater than is “$>$”. Therefore, $10 > 5$. If there is no sign in front of a number, it is assumed the number is a positive number.

Practice

Insert the correct mathematical symbol $>$ or $<$ for the following pairs of numbers. Check your answers with the answer key at the end of the book.

1. 5___11

2. 1___–2

3. –4___0

4. –2___–8

5. –35___–18

6. 0___–6

7. 12___0

8. 14___–23

► Adding and Subtracting Integers

You can add and subtract integers; but before you can do that, you need to know the rules for determining the sign of an answer. Of course, a calculator can determine the sign of your answer when you are working with integers. However, it is important that you know how to determine the sign of your answer without the use of a calculator. There are situations where you will be unable to use a calculator and will be handicapped if you do not know the rules for determining the sign of an answer. Knowing how to determine the sign of your answer is a basic algebra skill and is **absolutely necessary** to progress to more advanced algebra skills.

What Are the Sign Rules for Adding Integers?

When the signs of the numbers are the same, **add** the numbers and keep the same sign for your answer.

Examples: $-3 + -5 = -8$

$4 + 3 = 7$

If the signs of the numbers are different (one is positive and one is negative), then treat both of them as positive for a moment. Subtract the smaller one from the larger one, then give this answer the sign of the larger one.

Examples: $4 + {}^{-}7 = -3$

The answer is negative because 7 is bigger than 4 when we ignore signs.

Tip

If there is no sign in front of a number, the number is positive.

Practice

______ **9.**	$7 + 5$	______ **13.**	$-3 + 10$
______ **10.**	$-4 + {}^{-}8$	______ **14.**	$3 + {}^{-}9$
______ **11.**	$-17 + 9$	______ **15.**	$11 + {}^{-}2$
______ **12.**	$-9 + {}^{-}2$	______ **16.**	$-5 + 5$

What Are the Sign Rules for Subtracting Integers?

All subtraction problems can be converted to addition problems because subtracting is the same as adding the opposite. Once you have converted the subtraction problem to an addition problem, use the *Rules for Adding Integers* on the previous page.

For example, $2 - 5$ can also be written as $2 - {}^{+}5$. Subtraction is the same as adding the opposite, so $2 - {}^{+}5$ can be rewritten as $2 + {}^{-}5$. Because the problem has been rewritten as an addition problem, you can use the *Rules for Adding Integers*. As you recall, the rule says that if the signs are different, you should subtract the numbers and take the sign of the larger number and disregard the sign. Therefore, $2 + {}^{-}5 = -3$. See the following examples.

Examples: $7 - 3 = 7 - {}^{+}3 = 7 + {}^{-}3 = 4$

$6 - {}^{-}8 = 6 + {}^{+}8 = 14$

$-5 - {}^{-}11 = -5 + {}^{+}11 = 6$

Practice

______ **17.**	$5 - 6$	______ **23.**	$-8 - {}^{-}2$
______ **18.**	3 ____ -6	______ **24.**	$-11 - {}^{-}6$
______ **19.**	$-2 - 5$	______ **25.**	$10 - 3$
______ **20.**	$-7 - 12$	______ **26.**	$6 - {}^{-}6$
______ **21.**	$-9 - 3$	______ **27.**	$9 - 9$
______ **22.**	$-15 - {}^{-}2$	______ **28.**	$-8 - 10$

Shortcuts and Tips

Here are some tips that can shorten your work and save you time!

Tip #1: You may have discovered that $(- {}^{-})$ will be the same as a positive number. Whenever you have a problem with two negative signs side by side, change both signs to positive. Then work the problem.

Example: $4 - {}^{-}8 = 4 + {}^{+}8 = 4 + 8 = 12$

Tip #2: Notice that the subtraction sign is bigger and lower than the negative sign. You may have discovered that the subtraction sign gives you the same answer as a negative sign. You will find that the most frequently used notation is $5 - 9$ rather than $5 + {}^{-}9$.

Example: $3 - 5 = 3 + {}^{-}5 = {}^{-}2$ so $3 - 5 = {}^{-}2$

Tip #3: When adding more than two numbers, add all the positive numbers, add all the negative numbers, then add the resulting positive and negative numbers to obtain the answer.

Example: $2 + 3 + 5 - 7 = 10 - 7 = 3$

$-2 + 5 + 7 + {}^{-}6 = 12 + {}^{-}8 = 4$

$5 - 7 + 6 - 9 = 11 + {}^{-}16 = -5$

Practice

Practice using the shortcuts.

_______ **29.** $3 - {}^{-}2$

_______ **30.** $5 - 11$

_______ **31.** $5 - {}^{-}7$

_______ **32.** $9 - {}^{-}5$

_______ **33.** $-1 - 1$

_______ **34.** $-6 - 5$

_______ **35.** $-11 - {}^{-}12$

_______ **36.** $-7 - {}^{-}7$

_______ **37.** $13 - 5$

_______ **38.** $-8 - {}^{-}12$

_______ **39.** $4 + 11 + 5 - 10$

_______ **40.** $7 + 4 - 5 + 2$

_______ **41.** $-7 + 5 + 9 - 4$

_______ **42.** $12 - 3 + 4 + 6$

_______ **43.** $-9 - 11 - 2 + 5$

_______ **44.** $-8 + 12 - 5 + 6$

_______ **45.** $14 - 11 + 7 - 6$

_______ **46.** $-5 - {}^{-}7 + {}^{-}4 + 10$

_______ **47.** $-2 + 5 + 7 - 3 + 6$

_______ **48.** $3 - 10 - {}^{-}6 + 5 + {}^{-}7$

▶ Multiplying and Dividing Integers

When you are multiplying or dividing integers, if the signs are the same, the answer will be positive. If the signs are different, the answer will be negative. The dot (·) indicates multiplication in the following examples.

Examples: $2 \cdot 5 = 10$

$-2 \cdot {}^{-}5 = 10$

$-2 \cdot 5 = -10$

$10 \div {}^{-}2 = -5$

$-10 \div 2 = -5$

$-10 \div {}^{-}2 = 5$

Here's a tip to use when multiplying more than two numbers at a time.

Tip

If you are multiplying more than two numbers, use the odd-even rule. Count the number of negative signs in the problem. If the problem has an odd number of negative signs, the answer will be negative. If there is an even number of negative signs, the answer will be positive.

Examples: $2 \cdot 3 \cdot {}^{-}5 = -30$

$-5 \cdot 2 \cdot {}^{-}3 = 30$

$-7 \cdot 3 \cdot {}^{-}2 \cdot {}^{-}1 = -42$

Note: Zero is considered an even number.

Practice

______**49.** $7 \cdot 8$

______**50.** $-4 \cdot 5$

______**51.** $-14 \div 2$

______**52.** $-12 \cdot {}^{-}2$

______**53.** $-56 \cdot {}^{-}8$

______**54.** $-33 \div 3$

______**55.** $-44 \div {}^{-}11$

______**56.** $24 \div {}^{-}3$

______**57.** $7 \cdot {}^{-}11$

______**58.** $-75 \div {}^{-}3$

______**59.** $3 \cdot 2 \cdot {}^{-}4$

______**60.** $5 \cdot {}^{-}4 \cdot {}^{-}2$

______**61.** $-3 \cdot {}^{-}2 \cdot {}^{-}6$

______**62.** $3 \cdot 5 \cdot 6$

______**63.** $-2 \cdot 3 \cdot {}^{-}1 \cdot {}^{-}4$

______**64.** $-4 \cdot {}^{-}5 \cdot {}^{-}2 \cdot {}^{-}2$

Mixed Practice

Here is a mixture of problems for you to solve using what you've learned in this lesson. Work the problems without the use of a calculator.

_______ **65.** $2 - 7$

_______ **66.** $-2 \cdot 5$

_______ **67.** $-12 \div 4$

_______ **68.** $-9 + 12$

_______ **69.** $16 - {}^{-}5$

_______ **70.** $-36 \div {}^{-}18$

_______ **71.** $7 \cdot {}^{-}8$

_______ **72.** $-45 - 2$

_______ **73.** $-9 \cdot {}^{-}3$

_______ **74.** $-11 + {}^{-}15$

_______ **75.** $39 \div {}^{-}3$

_______ **76.** $-18 - {}^{-}3$

_______ **77.** $-20 \cdot 3$

_______ **78.** $-8 - {}^{-}8$

_______ **79.** $37 - 12$

_______ **80.** $-10 - 2$

_______ **81.** $-2 \cdot {}^{-}7 \cdot {}^{-}3$

_______ **82.** $2 + 9 - 11 + 3$

_______ **83.** $10 \cdot {}^{-}2 \cdot {}^{-}3$

_______ **84.** $-48 \div {}^{-}3$

_______ **85.** $20 - {}^{-}3 + 8 - 11$

_______ **86.** $2 \cdot {}^{-}2 \cdot {}^{-}4$

_______ **87.** $-10 - {}^{-}4$

_______ **88.** $60 \div {}^{-}5$

_______ **89.** $4 \cdot {}^{-}5 \cdot {}^{-}2 \cdot 10$

_______ **90.** $-10 + 5 - 7 - 13$

Applications

Represent the information in the problem with signed numbers. Then solve the problem.

Example: In Great Falls, Montana the temperature changed from 40 to 3 below zero. What was the difference in temperature?
$40° - {}^{-}3° = 43°$

91. The weather channel stated the high temperature for the day was 85° in Phoenix and the low temperature was 17° in Stanley, Idaho. What is the difference in the temperatures?

92. The predicted high for the day in Bismarck, North Dakota was 30° and the low was 7° below zero. What is the difference in the temperatures for the day?

93. You have a bank balance of $45 and write a check for $55. What is your new balance?

94. You have an overdraft of $20. You deposit $100. What is your new balance?

95. The water level in the town reservoir goes down 8 inches during a dry month, then gains 5 inches in a heavy rainstorm, and then loses another inch during the annual lawn sprinkler parade. What is the overall effect on the water level?

Skill Building until Next Time

When you balance your checkbook, you are working with positive and negative numbers. Deposits are positive numbers. Checks and service charges are negative numbers. Balance your checkbook using positive and negative numbers.

LESSON

Working with Algebraic Expressions

LESSON SUMMARY

In this lesson, you will use the same rules of signs that you learned in the previous lesson for any number—including fractions—not just integers. You will find out how to simplify and evaluate expressions and see how using the order of operations can help you find the correct answer.

▶ Simplifying Expressions

What does it mean when you are asked to *simplify* an expression? Numbers can be named in many different ways. For example: $\frac{1}{2}$, 0.5, 50%, and $\frac{3}{6}$ all name the same number. When you are told to simplify an expression, you want to get the simplest name possible. For example, because $\frac{3}{6}$ can be reduced, $\frac{1}{2}$ is the simplest name of the number.

Mathematical expressions, like numbers, can be named in different ways. For example, here are three ways to write the same expression:

1. $x + {}^{-}3$
2. $x + (-3)$ When you have two signs side by side, parentheses can be used to keep the signs separate.
3. $x - 3$ Remember that Lesson 1 showed that subtracting a positive 3 is the same as adding the opposite of a positive 3.

The operation of multiplication can be shown in many ways. In Lesson 1, we used the dot (·) to indicate multiplication. A graphics calculator will display an asterisk when it shows multiplication. You are probably familiar with this notation (2 × 3) to show multiplication. However, in algebra, we rarely use the × to indicate multiplication since it may be unclear whether the × is a variable or a multiplication sign. To avoid confusion over the use of the ×, we express multiplication in other ways. Another way to indicate multiplication is the use of parentheses (2)(3); also, when you see an expression such as $3ab$, it is telling you to multiply 3 times a times b.

Order of Operations

In order to simplify an expression that contains several different operations (addition, subtraction, multiplication, division, and such), it is important to do them in the right order. This is called the *order of operations.*

For example, suppose you were asked to simplify the following expression:

$$3 + 4 \cdot 2 + 5$$

At first glance, you might think it is easy: $3 + 4 = 7$ and $2 + 5 = 7$, then $7 \cdot 7 = 49$. Another person might say $3 + 4 = 7$ and $7 \cdot 2 = 14$ and $14 + 5 = 19$. Actually, both of these answers are wrong! To eliminate the possibility of getting several answers for the same problem, there is a specific order you must follow. This specific order is called *order of operations* and the steps you follow are:

1. Perform the operations inside grouping symbols such as ( ), { }, and [ ]. The division bar can also act as a grouping symbol. The division bar or fraction bar tells you to do the steps in the numerator and the denominator before you divide.
2. Evaluate exponents (powers) such as 3^2.
3. Do all multiplication and division *in order from left to right.*
4. Do all addition and subtraction *in order from left to right.*

Here are the steps for getting the correct answer to $3 + 4 \cdot 2 + 5$:

Multiply $4 \cdot 2$.	$3 + 4 \cdot 2 + 5$
Add the numbers in order from left to right.	$3 + \mathbf{8} + 5$
	$= 16$

The correct answer is 16.

Let's try some more examples.

Example: $3 + 4 \div 2 \cdot 3 + 5$

You need to do division and multiplication first *in order from left to right.* The division comes first, so first you divide and then you multiply.

Divide $4 \div 2$.	$3 + \mathbf{4} \div \mathbf{2} \cdot 3 + 5$
Multiply $2 \cdot 3$.	$3 + \mathbf{2} \cdot \mathbf{3} + 5$
Add the numbers in order from left to right.	$3 + \mathbf{6} + 5$
	$= 14$

Example: $7 + 3(6 - 2) - 6$

You need to perform the operation inside the ( ) first. Here, the parentheses indicate multiplication. The notation $3(6 - 2)$ means $3 \cdot (6 - 2)$.

Subtract $6 - 2$.	$7 + 3\,(6 - 2) - 6$
Multiply $3 \cdot 4$.	$7 + 3(\mathbf{4}) - 6$
Add and subtract the numbers in order from left to right.	$7 + \mathbf{12} - 6$
	$\mathbf{19} - 6$
	$= 13$

Example: $3 + 2^3 + 3(15 \div 3) - 2 + 1$

The expression contains parentheses ( ), so you do what's inside the ( ) first. Your next step is to simplify the exponent.

Divide $15 \div 3$.	$3 + 2^3 + 3(15 \div 3) - 2 + 1$
Simplify the exponent 2^3.	$3 + 2^3 + 3(\mathbf{5}) - 2 + 1$
Multiply $3 \cdot 5$.	$3 + \mathbf{8} + 3(5) - 2 + 1$
Add and subtract in order from left to right.	$3 + 8 + \mathbf{15} - 2 + 1$
	$= \mathbf{11} + 15 - 2 + 1$
	$= \mathbf{26} - 2 + 1$
	$= \mathbf{24} + 1$
	$= 25$

Tip

In order to remember that the order of operations is **P**arentheses, **E**xponents, **M**ultiplication & **D**ivision, then **A**ddition & **S**ubtraction, you can use the sentence: "**P**lease **e**xcuse **m**y **d**ear **A**unt **S**ally."

Practice

Use the order of operations to simplify the problems. Check your answers with the answer key at the end of the book.

1. $8 + 4 \cdot 2$

2. $2 + 3 - 4 + 5$

3. $5 - 3 \cdot 2^2$

4. $3 \cdot 10 - 18 \div 2$

5. $17 - 3 \cdot 4 \div 2$

6. $17 - 4 \div 2 \cdot 4$

7. $12 + 3^2 - 11$

8. $2 + 5(6 \div 3) + 4$

9. $45 \div 5 - 6$

10. $14 \div (3 + 2 \cdot 2)$

11. $5 + (7 - 5)^3 + 2$

12. $5 + 14 \div 7 + 5 \cdot 2$

13. $(9 - 5) - (6 + 1)$

14. $4(2 -5) + 8$

15. $15 \div 3 \cdot 2^3$

16. $4 + 2(^-6) + 15$

17. $3 - {}^-4 \cdot {}^-5$

18. $5 - 3(4)^2$

19. $2 + 12 \div 6 - 3 \cdot 2$

20. $3(2) \div 2(3) - 5$

Working with Multiple Grouping Symbols

What would you do if you had grouping symbols inside grouping symbols? To simplify the expression, $2\{4 + 3[10 - 4(2)] + 1\}$, start from the inside and work to the outside. Your first step is to multiply 4(2).

Subtract $10 - 8$.	$2\{4 + 3[10 - \mathbf{8}] + 1\}$
Multiply $3 \cdot 2$.	$2\{4 + 3[\mathbf{2}] + 1\}$
Add $4 + 6 + 1$.	$2\{4 + \mathbf{6} + 1\}$
Multiply $2 \cdot 11$	$2\{\mathbf{11}\}$
	$= 22$

Try this problem:

$2\{12 \div 2[10 - 2(2)] + 3\}$

If you got 8, you are correct!

► Evaluating Algebraic Expressions

What is the difference between simplifying an expression and evaluating an expression? In algebra, letters, called *variables*, are often used to represent numbers. When you are asked to *evaluate* an algebraic expression, you substitute a number in place of a variable (letter) and then simplify the expression. Study these examples.

Example: Evaluate the expression $2b + a$ when $a = 2$ and $b = 4$.

Substitute 2 for the variable a and 4 for the variable b. When the expression is written as $2b$, it means 2 times b. — $2b + a$

Multiply $2 \cdot 4$. — $2(\mathbf{4}) + \mathbf{2}$

$8 + 2$

$= 10$

Example: Evaluate the expression $a^2 + 2b + c$ when $a = 2$, $b = 3$, and $c = 7$.

Substitute 2 for a, 3 for b, and 7 for c. — $a^2 + 2b + c$

Find the value of 2^2. — $(\mathbf{2})^2 + 2(\mathbf{3}) + \mathbf{7}$

Multiply $2 \cdot 3$. — $\mathbf{4} + 2(3) + 7$

Add the numbers. — $4 + \mathbf{6} + 7$

$= 17$

Practice

Evaluate the algebraic expressions when $a = 2$, $b = -3$, $c = \frac{1}{2}$, $d = 7$, and $e = 4$.

21. $3a + e$

22. $ac + d$

23. $4cd$

24. $bd - 3e$

25. $a^3 + ce$

26. $5ab - d$

27. $-ab + 10$

28. $4b - 5e$

29. $3e^2 + 4b$

30. $3(a + b + d)$

31. $c(12a)$

32. $c^2e + a$

33. $d^2 + 7b$

34. $5(a + b)^2$

35. $6ce + 4a - d$

36. $5a^4$

37. $ba^3 + d^2$

38. $(e + d)(e - d)$

39. $9(a + d) - 9^2$

40. $6(d + 3b) + 7ace$

Skill Building until Next Time

The next time you go shopping, take note of the price on any two items. Use a variable to represent the cost of the first item and a different variable to represent the cost of the second item. Use the variables to write an algebraic expression that will calculate what you spent on the combination of the two items. Evaluate the expression to answer the problem.

LESSON

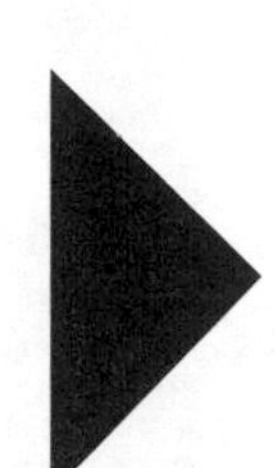

Combining Like Terms

LESSON SUMMARY

In this lesson, you will simplify expressions by combining like terms. You will need this skill in future lessons to solve algebraic equations. Also, you will find out how to use the distributive property to help you simplify expressions.

▶ What Are Like Terms?

First of all, what are terms? *Terms* are connected by addition or subtraction signs. For example: The expression $a + b$ has two terms and the expression ab has one term. Remember, ab means a times b. The expression ab is one term because it is connected with an understood multiplication sign. The expression $3a + b + 2c$ has three terms. The expression $3ab + 2c$ has two terms.

Secondly, what are like terms and why are they important? *Like terms* have the same variable(s) with the same exponent, such as $3x$ and $10x$. More examples of like terms are:

$3ab$ and $7ab$

$2x^2$ and $8x^2$

$4ab^2$ and $6ab^2$

5 and 9

You can add and subtract like terms. When you add and subtract like terms, you are simplifying an algebraic expression. How do you add like terms? Simply **add the numbers in front of the variables and keep the variables the**

same. The numbers in front of the variables are called *coefficients*. Therefore, in the expression $6x + 5$, the coefficient is 6. Here are some sample problems.

Example: $2x + 3x + 7x$

Add the numbers in front of the variables.	$2 + 3 + 7$
Don't change the variable.	$= 12x$

Example: $4xy + 3xy$

Add the numbers in front of the variables.	$4 + 3$
Don't change the variables.	$= 7xy$

Example: $2x^2y - 5x^2y$

Subtract the numbers in front of the variables.	$2 - 5$
Don't change the variables.	$= -3x^2y$

Example: $4x + 2y + 9 + 6x + 2$

(Hint: You can only add the like terms, $4x$ and $6x$, and the numbers 9 and 2.)

Add the like terms.	$4x + 2y + 9 + 6x + 2$
	$= 10x + 2y + 11$

Practice

Simplify the expressions by combining like terms.

______ **1.** $2x + 4x + 11x$

______ **2.** $6x + 3y + 4y + 7x$

______ **3.** $6x + 5 + 9x$

______ **4.** $12x - 3x + 9x$

______ **5.** $5 - 4x + 2x^2 + 7 - x$

______ **6.** $9x - 2y - 5y + 2 - 4x$

______ **7.** $12x + 12y + 5$

______ **8.** $5x^2 + 7x + 5 + 3x^2 - 2x + 4$

______ **9.** $-12x + 12x + 8$

______ **10.** $3xy + 5x - 2y + 4yx + 11x - 8$

► Using the Distributive Property to Combine Like Terms

What do you do with a problem like this: $2(x + y) + 3(x + 2y)$? According to the *order of operations* that you learned in the previous lesson, you would have to do the grouping symbols first. However, you know you can't add x to y because they are not like terms. What you need to do is use the *distributive property*. The distributive property tells you to multiply the number and/or variable(s) outside the parentheses by every term inside the parentheses. You would work the problem like this:

Multiply 2 times x and 2 times y. Then multiply 3 times x and 3 times $2y$. If there is no number in front of the variable, it is understood to be 1, so 2 times x means 2 times $1x$. To multiply, you multiply the numbers and the variable stays the same. When you multiply 3 times $2y$, you multiply 3 times 2 and the variable, y, stays the same, so you would get $6y$. After you have multiplied, you can then combine like terms.

Example:

Multiply $2(x + y)$ and $3(x + 2y)$. $2(x + y) + 3(x + 2y)$

Combine like terms. $2x + 2y + 3x + 6y$

$= 5x + 8y$

Tip

If there is no number in front of a variable, it is understood to be 1.

Here are two more examples using the distributive property.

Example: $2(x + y) + 3(x - y)$

Multiply 2 times x and 2 times y. Then multiply 3 times x and 3 times $(-y)$. When you multiply 3 times $(-y)$, this is the same as $3(-1y)$. The 1 is understood to be in front of the y even though you don't see it. In this example, you can see how the parentheses are used to indicate multiplication. $2(x + y) + 3(x - y)$

Use the distributive property. $2x + 2y + 3x - 3y$

Combine like terms. $= 5x - y$

Example: $2(2x + y) - 3(x + 2y)$

Use the distributive property to get rid of the parentheses. The subtraction sign in front of the 3 is the same as multiplying $(-3)(x)$ and $(-3)(2y)$.

Use the distributive property. $2(2x + y) - 3(x + 2y)$

Combine like terms. $4x + 2y - 3x - 6y$

$= x - 4y$

Practice

Use the distributive property to simplify the expressions.

______**11.** $4x + 2(x + y)$

______**12.** $2(a + b) + 6(a + b)$

______**13.** $5(m + 2) + 3(m + 7)$

______**14.** $5(x - 2y) + 2(y - x)$

______**15.** $8(r + s) - 5s + 4r$

______**16.** $3(m + n) + 4(2m - 3) + 2(5 - n)$

_______**17.** $3(x - y) + 5(2x + 3y)$

_______**18.** $5 - 3(x + 7) + 8x$

_______**19.** $-5(a + 3b) - 2(4a - 4b)$

_______**20.** $12 + 3(x + 2y - 4) - 7(-2x - y)$

Mixed Practice

Here's a chance to do some review of the skills you have learned so far. Simplify the expressions.

_______**21.** $-2(3) + 8 - 10$

_______**22.** $24 - 11 + 5 - 16$

_______**23.** $-8 - {}^{-}5 + (1 + 2)^2$

_______**24.** $5(-2) - 3(4)$

_______**25.** $8 + \frac{16}{-2}$

_______**26.** $-10(\frac{1}{10})$

_______**27.** $8 \div (-2) \cdot 5 + 2$

_______**28.** $\frac{-10}{2} + \frac{20}{4}$

_______**29.** $(5)(-2)(-3)(-1)$

_______**30.** $7(-2)(0)(-5)(4)$

Practice

Evaluate the following algebraic expressions when $a = \frac{1}{2}$, $b = -3$, $c = 6$, $d = 12$, and $e = 2$.

_______**31.** abc

_______**32.** $2a + 3c$

_______**33.** $2(a + b)$

_______**34.** $a(2b + c)$

_______**35.** $3(b + d) - 4a$

_______**36.** $2c^2 + 4e$

_______**37.** $-(\frac{d}{b} + 5e)$

_______**38.** $5(b + e)^2$

Skill Building until Next Time

To calculate your earnings for a week of work, you multiply the hours worked times the hourly wage. Write down how many hours you worked last week. At the end of this week, write down how many hours you worked this week. Use a variable to represent the hourly wage and write an algebraic expression to calculate your paycheck for the two-week period. Simplify the expression.

LESSON

Solving Basic Equations

LESSON SUMMARY

This lesson teaches the algebra skills needed to solve basic equations. You will learn how to solve equations using addition, subtraction, multiplication, and division. You will find out how to check your answers to make sure they are correct and find out ways to set up equations for solving word problems.

▶ What Is an Equation?

An equation is a mathematical tool that helps people solve many real-life problems. What is an equation? An *equation* is two equal expressions. These expressions could be numbers such as $6 = 5 + 1$ or variable expressions such as $D = rt$. What does it mean to solve an equation? When you find the value of the variable, you have solved the equation. For example, you have solved the equation $2x = 10$ when you know the value of x.

The basic rule for solving equations is: When you do something to one side of an equation, you must do the same thing to the other side of the equation. You'll know that you have solved an equation when the variable is alone (isolated) on one side of the equation and the variable is positive. In the example $-x = 5$, 5 is not the answer because the x variable is negative.

► Solving Equations Using Addition or Subtraction

You can solve equations using addition and subtraction by getting the variable on a side by itself. Think about how you would solve this equation: $x + 4 = 10$. Your goal is to get the variable x on a side by itself. How do you do that? To get x by itself, you need to get rid of the 4. How? If you subtract 4 from 4, the result is 0, and you have eliminated the 4 to get x on a side by itself. However, if you subtract 4 from the left side of the equation, then you must do the same to the right side of the equation. You would solve the equation this way:

Example: $x + 4 = 10$

Subtract 4 from both sides of the equation.	$x + 4 - 4 = 10 - 4$
Simplify both sides of the equation.	$x + 0 = 6$
Add 0 to x.	$x = 6$

When you add zero to a number, the number does not change, so $x + 0 = x$. When you do this, you are using the *additive property of zero,* which states that a number added to zero equals that number. For example, $5 + 0 = 5$.

Let's look at another equation: $x - 5 = 9$. What do you need to do to get the variable on a side by itself? You need to get rid of the 5. The equation says to subtract 5, so what undoes subtraction? If you said addition, you are right! In this problem, you need to add 5 to both sides of the equation to get rid of the 5.

Example: $x - 5 = 9$

Add 5 to both sides of the equation.	$x - 5 + 5 = 9 + 5$
Simplify both sides of the equation.	$x + 0 = 14$
Add 0 to x.	$x = 14$

Example: $a + 6 = 7$

Subtract 6 from both sides of the equation.	$a + 6 - 6 = 7 - 6$
Simplify both sides of the equation.	$a + 0 = 1$
Add 0 to a.	$a = 1$

Example: $y - 11 = 8$

Add 11 to both sides of the equation.	$y - 11 + 11 = 8 + 11$
Simplify both sides of the equation.	$y + 0 = 19$
Add 0 to y.	$y = 19$

Example: $-r + 9 = 13$

Subtract 9 from both sides of the equation.	$-r + 9 - 9 = 13 - 9$
Simplify both sides of the equation.	$-r + 0 = 4$
Add 0 to $-r$.	$-r = 4$

Are you finished? No! The value of the variable must be positive, and r is negative. You need to make the variable positive. You can do that by multiplying both sides of the equation by (-1).

$$(-1)(-r) = 4(-1)$$

$$r = -4$$

Practice

Solve these equations without the use of a calculator. If you get a fraction for an answer, reduce the fraction to simplest form.

_____ **1.** $a + 10 = 4$

_____ **2.** $d - 7 = 8$

_____ **3.** $x - 5 = -6$

_____ **4.** $y - {}^{-}10 = 3$

_____ **5.** $t - 8 = -3$

_____ **6.** $s - (-2) = 7$

_____ **7.** $-m + 5 = 1$

_____ **8.** $-b + 10 = 2$

_____ **9.** $-x + 12 = -5$

_____ **10.** $c - {}^{-}12 = -5$

► Checking Your Answers

You can use an answer key at the end of the book to check your answers. However, in the world of work there is no answer key to tell you if you got the correct answer to an equation. There is a way to check your answer. When you replace the variable with your answer, you should get a true statement. If you get a false statement, then you do not have the right answer. To check practice problem 1 in this lesson, you would do the following steps.

Check for Problem 1: $a + 10 = 4$

The answer you got for problem 1 was –6, so replace the variable, which is *a*, with –6. This is called *substitution*. You are substituting the variable with the number –6.

Substitute –6 in place of the variable, *a*. $-6 + 10 = 4$

Simplify the left side of the equation. $4 = 4$

Your result, 4 = 4, is a true statement, which tells you that you solved the equation correctly.

Check for Problem 2: $d - 7 = 8$

Substitute 15, your answer, in place of the variable. $15 - 7 = 8$

Simplify the left side of the equation. $8 = 8$

Because the result is a true statement, you know you solved the equation correctly.

Practice

Check problems 3–10.

▶ Solving Equations Using Multiplication and Division

In the equation $x + 10 = 2$, to get rid of the 10, you would subtract 10, which is the opposite of adding 10. In the equation $x - 5 = 6$, to get rid of the 5, you would add 5, which is the opposite of subtracting 5. So in the equation $5x = 10$, how do you think you would get rid of the 5? What is the opposite of multiplying by 5? Yes, the opposite of multiplying by 5 is dividing by 5. You would solve these equations using division.

$5x = 10$	$4x = 12$
$\frac{5x}{5} = \frac{10}{5}$	$\frac{4x}{4} = \frac{12}{4}$
$x = 2$	$x = 3$

When you divide a number by itself, you always get 1, so $5x$ divided by 5 equals $1x$. Remember that you learned in Lesson 3 that $1x$ is the same as x.

In the equation $\frac{x}{5} = 2$, how would you get rid of the 5 so the x will be on a side by itself? If you said multiply by 5, you are correct! The opposite of dividing by 5 is multiplying by 5. When you multiply $\frac{1}{5} \cdot 5$, you get 1. If you multiply $\frac{x}{5}$ times 5, you will get $1x$. Remember that the 1 in front of the x is understood to be there when there is no number in front of x. Study these examples:

$\frac{x}{2} = 13$	$\frac{x}{5} = 3$
$\frac{x}{2} \cdot 2 = 13 \cdot 2$	$\frac{x}{5} \cdot 5 = 3 \cdot 5$
$x = 26$	$x = 15$

Practice

Solve these equations without the use of a calculator. If you get a fraction for an answer, reduce the fraction to its simplest form.

______ **11.** $2x = 24$

______ **12.** $\frac{x}{2} = 24$

______ **13.** $3x = 27$

______ **14.** $\frac{x}{8} = 16$

______ **15.** $\frac{x}{13} = 2$

______ **16.** $3x = 4$

______ **17.** $6x = 10$

______ **18.** $\frac{x}{3} = \frac{2}{3}$

______ **19.** $\frac{x}{2} = \frac{2}{3}$

______ **20.** $45x = 15$

Mixed Practice

Solve the equations without the use of a calculator. If you get a fraction for an answer, reduce the fraction to its simplest form.

_____**21.** $x + 3 = 11$

_____**22.** $x - \frac{1}{2} = 4$

_____**23.** $3x = 10$

_____**24.** $\frac{x}{7} = 3$

_____**25.** $16x = 48$

_____**26.** $x - {}^{-}12 = -3$

_____**27.** $x + 23 = 5$

_____**28.** $\frac{x}{14} = 3$

_____**29.** $x + 4 = -10$

_____**30.** $x + 100 = 25$

_____**31.** $4x = -32$

_____**32.** $\frac{x}{9} = -7$

_____**33.** $-3x = 39$

_____**34.** $x + 17 = 11$

_____**35.** $-\frac{s}{6} = -5$

_____**36.** $-x - 11 = 18$

_____**37.** $5x = \frac{1}{2}$

_____**38.** $-8x = 6$

_____**39.** $x + 2.3 = -4.7$

_____**40.** $.3x = -12$

▶ Setting Up Equations for Word Problems

Equations can be used to solve real-life problems. In an equation, the variable often represents the answer to a real-life problem. For example, suppose you know that you can earn twice as much money this summer as you did last summer. You made $1,200 last summer. How much will you earn this summer? You can use the variable x to represent the answer to the problem. You want to know how much you can earn this summer.

Let x = how much you can earn this summer

$1,200 = amount earned last summer

$x = 2 \cdot 1{,}200$

$x = \$2{,}400$

You might be wondering why you should use algebra to solve a problem when the answer can be found using arithmetic. Good question! If you practice using equations for simple problems, then you will find it easier to write equations for problems that can't be solved using arithmetic.

Example: The cost of a meal, including a 20% tip, is $21.60. How much was the meal without the tip?

Let x = cost of the meal without the tip

The tip is 20% of x or $0.2x$

Together, $x + 0.2x = 21.60$

$1.2x = 21.6$

$x = 18$

Without the tip, the meal cost $18.

Example: There are twice as many girls in your algebra class as there are boys. If there are 18 girls in the class, how many boys are in the class?

Let x = the number of boys in the class.

Then 2 times the number of boys is equal to the number of girls. You can represent that in the equation $2x = 18$.

$x = 9$; therefore, there are 9 boys in the class.

Example: You are going to be working for Success Corporation. You got a signing bonus of $2,500, and you will be paid $17.50 an hour. If you are paid monthly, how much will your first paycheck be? Be sure to include your signing bonus, and assume that you have a 40-hour work week and there are 4 weeks in this month.

Let x = first monthly paycheck

Then $x = 4 \cdot 40 \cdot \$17.50 + \$2{,}500$

$x = \$5{,}300$

Practice

Use algebra to solve the following problems. You want to practice writing equations to solve the problems.

41. You have been contracted to replace a countertop in a kitchen. You will get paid $1,000 to do the job. You can complete the job in three days working 8 hours a day. You will need to pay the cost of the materials, which is $525, with the $1,000 that you receive to do the job. How much will you be paid an hour for your time?

42. How much is a 15% tip on a $32 meal?

43. When Kim works 40 hours in a week, she earns $560. If she only works 35 hours, how much should she expect to earn?

44. There are 40 questions on a test. How many questions do you need to have correct to score a 90% on the test?

45. Suppose you scored 80% on a test and answered 32 questions correctly. How many questions were on the test?

46. Susan and Jayne have part-time jobs selling tickets at the theatre. This week, Jayne's paycheck was twice the amount of Susan's paycheck. They both were paid $7.50 an hour, and Jayne's paycheck was $60. How many hours did Susan work this week? (Susan earns half the amount Jayne earns.)

47. You are going to plant a garden. The length of the garden is twice as long as the width. The perimeter of the lot is 30. (The perimeter of a rectangle = 2 lengths + 2 widths.) What are the dimensions of the length and the width?

48. Your starting salary as a beginning teacher in Castleford, Idaho, is $22,000. If it is recommended that you spend no more than 25% of your income on housing, how much should you be spending on housing? How much can you spend per month?

49. The cost of 4 pillows, plus 5% sales tax, was $46.20. How much did each pillow cost?

50. How fast should you drive if you want to travel 400 miles in 6 hours? (The distance formula is $D = rt$.)

Skill Building until Next Time

When you buy a new or used car, you may have to pay sales tax in addition to the price of the car. Go to a car lot or browse the newspaper for a vehicle that you would be interested in buying. If the sales tax is 5%, calculate the sales tax on the vehicle and then determine the total cost of the car. Begin by setting up an equation.

LESSON

Solving Multi-Step Equations

LESSON SUMMARY

In this lesson, you will use addition, subtraction, multiplication, and division to solve equations that have more than one step. You will also solve equations where the coefficient of the variable is a fraction.

▶ Solving Equations Requiring More Than One Step

You learned the basic steps needed to solve algebraic equations in the previous lesson, so now you're ready to solve more complicated equations. These equations may require two or more steps. Remember that the basic rules for solving equations are:

1. What you do to one side of the equation, you must do to the other side.
2. You can add, subtract, multiply, or divide both sides of an equation by the same number.

Can you guess what steps you would use to solve $3x + 1 = 10$? Ask yourself which numbers you need to get rid of so that the variable will be on a side by itself. In other words, you need to isolate the variable. To get the x on a side by itself or to isolate the variable, you will need to eliminate the 3 and the 1. Does it matter which number you get rid of first? Yes! You generally get rid of the number with the variable—that is, the coefficient of x—last. Therefore, you would eliminate the 1 first and the 3 last.

Example: $3x + 1 = 10$

Subtract 1 from both sides of the equation.	$3x + 1 - 1 = 10 - 1$
Simplify both sides of the equation.	$3x = 9$
Divide both sides of the equation by 3.	$\frac{3x}{3} = \frac{9}{3}$
Simplify both sides of the equation.	$x = 3$

Example: $5x - 2 = 13$

Add 2 to both sides of the equation.	$5x - 2 + 2 = 13 + 2$
Simplify both sides of the equation.	$5x = 15$
Divide both sides of the equation by 5.	$\frac{5x}{5} = \frac{15}{5}$
Simplify both sides of the equation.	$x = 3$

Example: $\frac{x}{9} + 4 = 13$

Subtract 4 from both sides of the equation.	$\frac{x}{9} + 4 - 4 = 13 - 4$
Simplify both sides of the equation.	$\frac{x}{9} = 9$
Multiply both sides of the equation by 9.	$\frac{x}{9} \cdot 9 = 9 \cdot 9$
Simplify both sides of the equation.	$x = 81$

Example: $18 = 9 - 3x$

Subtract 9 from both sides of the equation.	$18 - 9 = 9 - 9 - 3x$
Simplify both sides of the equation.	$9 = -3x$
Divide both sides of the equation by –3.	$\frac{9}{-3} = \frac{-3x}{-3}$
Simplify both sides of the equation.	$-3 = x$

You know that you have solved an equation when you get x on a side by itself. The variable can be on the left side of the equation or on the right side of the equation. The equations $x = 3$ and $3 = x$ are the same, and both are acceptable answers.

Practice

Solve the equations. Show each step you use to solve the equation.

_____ **1.** $24x + 1 = 25$

_____ **2.** $3x + 4 = 19$

_____ **3.** $13x - 11 = 15$

_____ **4.** $2x - 3 = -13$

_____ **5.** $\frac{x}{3} + 2 = 3$

_____ **6.** $\frac{x}{7} - 7 = 2$

_____ **7.** $-2x + 3 = 17$

_____ **8.** $-2x - 3 = 2$

_____ **9.** $2 = 5x + 12$

_____ **10.** $-10 = 3x + 5$

_____ **11.** $-13 = -3x + 2$

_____ **12.** $-\frac{x}{3} + 2 = 4$

_____ **13.** $3x + 1 = 6$

_____ **14.** $\frac{x}{4} - 4 = -14$

_____ **15.** $2.2x + 1 = 3.2$

_____ **16.** $\frac{x}{10} + 1.7 = -3.5$

_____ **17.** $-2x + 1 = 27$

_____ **18.** $3 - 2x = 1$

_____ **19.** $5 - \frac{x}{2} = 7$

_____ **20.** $5x + 2 = 2$

Setting Up Equations for Word Problems

You can use the same rules discussed at the beginning of this lesson to figure out the correct answer to word problems. Take a look at the following examples.

Example: You received a raise in your hourly rate of pay. Your new rate of pay is $10 an hour. You are now earning $4 more than $\frac{2}{3}$ your original wage. How much did you make before you got your raise? What was your raise?

Let x = the original rate of pay.

Then $4 + \frac{2}{3}x = \$10$.

Solution: $4 + \frac{2}{3}x = 10$

Subtract 4 from both sides of the equation. $4 - 4 + \frac{2}{3}x = 10 - 4$

Simplify both sides of the equation. $\frac{2}{3}x = 6$

Multiply both sides of the equation by $\frac{3}{2}$. $\frac{2}{3}x \cdot \frac{3}{2} = 6 \cdot \frac{3}{2}$

Simplify both sides of the equation. $x = \$9$

Your raise was $\$10 - \$9 = \$1$.

Example: You work at a gift shop and have a 20% employee discount on any regularly priced merchandise you buy. You found a sale-priced item for \$14. You have \$46 to spend for gifts. How much can you spend on regularly priced items?

Let x = regularly priced merchandise.

If you receive a discount of 20%, then you are paying 80% of the regular price.

Solution:	$\$14 + .80x = \$46$
Subtract 14 from both sides of the equation.	$14 - 14 + .80x = 46 - 14$
Simplify both sides of the equation.	$.80x = 32$
Divide both sides of the equation by .80.	$\frac{.80}{.80}x = \frac{32}{.80}$
Simplify both sides of the equation.	$x = \$40$

21. Jack earns $\frac{2}{3}$ as much as Jill. His yearly income is \$38,000. How much does Jill earn?

22. You can save $\frac{1}{5}$ the cost of a new washing machine by paying cash. The sale price of the washing machine you plan to buy is \$400. What is the regular price of the washing machine? (If your savings are $\frac{1}{5}$ of the original cost, then your sale price must be $\frac{4}{5}$ the original cost.)

23. You have an exercise regimen that includes doing sit-ups. You are currently doing 5 more than twice the number of sit-ups that you did when you started your exercise program. You can currently do 35 sit-ups. How many did you do when you started exercising?

24. In Crissy's class, there are 3 more than twice as many girls as boys. If the whole class has 27 students, how many boys are there?

25. Suppose you just deposited \$500 into a bank account that has earned 5% interest on the amount you initially invested. If the new balance is \$1,760, what was your initial investment?

▶ Solving Equations That Have a Fraction in Front of the Variable

How do you eliminate the number in front of the variable (coefficient) when it is a fraction? There are many approaches you could use to solve an equation with a fraction for a coefficient, but let's use the shortest method. The shortest method is to multiply both sides of the equation by something called the *multiplicative inverse* of the coefficient. You're probably wondering, "What is the multiplicative inverse of a number?" Take a look at these examples: The multiplicative inverse of 2, which can be written as $\frac{2}{1}$, is $\frac{1}{2}$. The multiplicative inverse of 5, which can be written as $\frac{5}{1}$, is $\frac{1}{5}$. The multiplicative inverse of $\frac{2}{3}$ is $\frac{3}{2}$. The multiplicative inverse of $-\frac{3}{4}$ is $-\frac{4}{3}$. You probably get the idea now—you get the multiplicative inverse of a number by inverting the number. In other words, by turning it upside down!

A number times its multiplicative inverse will always equal 1. Thus, $\frac{4}{5} \cdot \frac{5}{4} = 1$. Here's how you use the multiplicative inverse to solve equations with a fraction in front of the variable:

Example: $\frac{2}{3}x + 1 = 5$

Subtract 1 from both sides of the equation.	$\frac{2}{3}x + 1 - 1 = 5 - 1$
Simplify both sides of the equation.	$\frac{2}{3}x = 4$
Multiply both sides of the equation by the multiplicative inverse of the coefficient.	$\frac{2}{3}x \cdot \frac{3}{2} = 4 \cdot \frac{3}{2}$
Simplify both sides of the equation.	$x = 6$

Example: $\frac{3}{5}x - 2 = 7$

Add 2 to both sides of the equation.	$\frac{3}{5}x - 2 + 2 = 7 + 2$
Simplify both sides of the equation.	$\frac{3}{5}x = 9$
Multiply both sides of the equation by $\frac{5}{3}$.	$\frac{3}{5}x \cdot \frac{5}{3} = 9 \cdot \frac{5}{3}$
Simplify both sides of the equation.	$x = 15$

Practice

Solve the equations. Show all the steps needed to solve each equation.

______ **26.** $\frac{4}{7}x - 1 = 11$

______ **27.** $\frac{5}{6}x + 4 = 14$

______ **28.** $-\frac{3}{4}x + 5 = 2$

______ **29.** $7 = \frac{1}{2}x + 2$

______ **30.** $36 = -\frac{4}{3}x + 4$

______ **31.** $\frac{17}{2}x - 2 = 15$

______ **32.** $\frac{5}{8}x + 3 = 3$

______ **33.** $-\frac{4}{5}x - 14 = -2$

______ **34.** $\frac{6}{5}x + 2 = 7$

______ **35.** $-\frac{2}{3}x - 1 = 2$

Skill Building until Next Time

It is recommended that you spend no more than 25% of your income on housing. Calculate the recommended housing expenditure based on your income or your parents' income. Are you within the guideline? What other expenses need to be included when calculating one's housing expense other than the rent or mortgage payment?

LESSON

Solving Equations with Variables on Both Sides of the Equation

LESSON SUMMARY

You know that when you solve equations, the object is to get the variable on a side by itself or, in other words, isolate the variable. In this lesson, you will solve equations that have variables on both sides of the equation.

▶ What Do You Do When You Have Variables on Both Sides of the Equation?

To solve the equation, $2x + 5 = x - 3$, you can start by using the techniques you already know. You know that you want to isolate the variable on a side by itself. You don't want the 5 with the variable $2x$. Get rid of the 5 by subtracting 5 from both sides of the equation.

Example: $2x + 5 = x - 3$

Subtract 5 from both sides of the equation. $2x + 5 - 5 = x - 3 - 5$

Simplify both sides of the equation. $2x = x - 8$

You still don't have the variable isolated because you have variables on both sides of the equation. You need to get rid of the x on the right side of the equation, and you know that $x - x = 0$, so subtract x from both sides of the equation.

Subtract x from both sides of the equation. $2x - x = x - x - 8$

If there is no number (coefficient) in front of the x, it is understood to be 1. So $2x$ minus $1x$ is $1x$ or x. And $1x$ minus $1x$ is zero.

Simplify both sides of the equation to get your answer. $x = -8$

Example: $3x - 2 = 2x + 5$

Add 2 to both sides of the equation.	$3x - 2 + 2 = 2x + 5 + 2$
Simplify both sides of the equation.	$3x = 2x + 7$
Subtract $2x$ from both sides of the equation.	$3x - 2x = 2x - 2x + 7$
Simplify both sides of the equation.	$x = 7$

Example: $6x - 5 = 3x + 4$

Add 5 to both sides of the equation.	$6x - 5 + 5 = 3x + 4 + 5$
Simplify both sides of the equation.	$6x = 3x + 9$
Subtract $3x$ from both sides of the equation.	$6x - 3x = 3x - 3x + 9$
Simplify both sides of the equation.	$3x = 9$
Divide both sides of the equation by 3.	$\frac{3x}{3} = \frac{9}{3}$
Simplify both sides of the equation.	$x = 3$

Practice

You can see that you are starting to work more complex equations. Be sure to write out all steps. This may seem like unnecessary work, but taking the time to write out all the steps actually saves you time. You will be less apt to make a mistake. Also, it is easier to find and correct a mistake if you write out your steps.

______ **1.** $7x + 2 = 3x - 6$

______ **2.** $5x - 3 = 3x + 7$

______ **3.** $8x + 2 = 4x - 6$

______ **4.** $9x + 4 = 8x + 12$

______ **5.** $9x + 12 = 6x + 3$

______ **6.** $6x + 4 = 2x + 12$

______ **7.** $8x + 7 = 2x + 3$

______ **8.** $4x - 5 = 5x + 1$

______ **9.** $2x + 5 = 5x - 1$

______ **10.** $3.5x + 1 = 3x + 4$

► Using the Distributive Property

To solve some equations with variables on both sides of the equation, you will need to use the distributive property that you learned about in Lesson 3. The distributive property tells you to multiply the number outside the parentheses by each term inside the parentheses in equations like $2(x + 2) = 2x + 4$ and $3(a - b) = 3a - 3b$.

Example: $5x + 3 = 3(x + 5)$

Use the distributive property.	$5x + 3 = 3x + 15$
Subtract 3 from both sides of the equation.	$5x + 3 - 3 = 3x + 15 - 3$
Simplify both sides of the equation.	$5x = 3x + 12$
Subtract $3x$ from both sides of the equation.	$5x - 3x = 3x - 3x + 12$
Simplify both sides of the equation.	$2x = 12$
Divide both sides of the equation by 2.	$\frac{2x}{2} = \frac{12}{2}$
Simplify both sides of the equation.	$x = 6$

Example: $4x + 6 = -2(3x + 4)$

Use the distributive property.	$4x + 6 = -6x - 8$
Subtract 6 from both sides of the equation.	$4x + 6 - 6 = -6x - 8 - 6$
Simplify both sides of the equation.	$4x = -6x - 14$
Add $6x$ to both sides of the equation.	$4x + 6x = -6x + 6x - 14$
Simplify both sides of the equation.	$10x = -14$
Divide both sides of the equation by 10.	$\frac{10x}{10} = -\frac{14}{10}$
Simplify both sides of the equation.	$x = -\frac{14}{10} = -\frac{7}{5} = -1\frac{2}{5}$

Practice

Be sure to write out your steps!

_______ **11.** $4x + 2 = 2(x + 3)$

_______ **12.** $3x + 3 = 2(x - 3)$

_______ **13.** $4x + 6 = 2(3x - 4)$

_______ **14.** $6x - 2 = 2(2x + 3)$

_______ **15.** $6x - 3 = 3(3x - 5)$

_______ **16.** $4.2x + 6 = 2(2x + 3)$

_______ **17.** $11x - 5 = 7(x - 2)$

_______ **18.** $8x + 5 = -3(x + 2)$

_______ **19.** $4x + 12 = 3(4 - x)$

_______ **20.** $11x - 3 = -3(-x + 3)$

▶ Solving More Complex Equations

In this section, you will use the distributive property and combine similar terms to simplify equations. The equations will also contain variables on both sides of the equation. Look at the following examples very carefully.

Example: $2x + 3 + 3x = 4x - 7 + 9$

Combine similar terms on both sides of the equation.	$5x + 3 = 4x + 2$
Subtract 3 from both sides of the equation.	$5x + 3 - 3 = 4x + 2 - 3$
Simplify both sides of the equation.	$5x = 4x - 1$
Subtract $4x$ from both sides of the equation.	$5x - 4x = 4x - 4x - 1$
Simplify both sides of the equation.	$x = -1$

Example: $5x + 3 - 2x = 2(x - 3) + 5$

Use the distributive property.	$5x + 3 - 2x = 2x - 6 + 5$
Combine similar terms on both sides of the equation.	$3x + 3 = 2x - 1$
Subtract 3 from both sides of the equation.	$3x + 3 - 3 = 2x - 1 - 3$
Simplify both sides of the equation.	$3x = 2x - 4$
Subtract $2x$ from both sides of the equation.	$3x - 2x = 2x - 2x - 4$
Simplify both sides of the equation.	$x = -4$

Practice

Write out all your steps.

_______ **21.** $4x - 3 + 3x = -2x + 5 - 17$

_______ **22.** $12x + 2 - 5x = 3(x + 5) + 3$

_______ **23.** $12x - 11x + 5 = -2(x + 3) + 5$

_______ **24.** $5 - (x - 3) = 10 - 3x$

_______ **25.** $3x + 2 - 6x = 8 + 3(x + 2)$

_______ **26.** $9x - (x + 3) = 2(x + 4) + 7$

_______ **27.** $5(2x + 3) = 4(x - 1) + 1$

_______ **28.** $1.2x + 2(x + .6) = 5x - 9.8 - 4x$

_______ **29.** $13x - 2(3x - 4) = 6(x + 2) - 4$

_______ **30.** $13x - 3(-2x - 3) = 4(x + 4) - 12$

► Equations Without a Variable in the Answer

In some equations, all the variables will be eliminated. These are special cases, and the equations may or may not have solutions.

Case 1

	$3x + 15 = 3(x + 10)$
Use the distributive property.	$3x + 15 = 3x + 30$
Subtract 15 from both sides of the equation.	$3x + 15 - 15 = 3x + 30 - 15$
Simplify both sides of the equation.	$3x = 3x + 15$
Subtract $3x$ from both sides of the equation.	$3x - 3x = 3x - 3x + 15$
Simplify both sides of the equation.	$0 = 15$

Your common sense tells you that 0 will never equal 15. This means there is no value of the variable that will make the equation true because 0 will never equal 15. Since there is no value of x that will ever make the equation true, there is no solution. When there is no solution, it is called an *empty set.* This notation $\varnothing$ is used for the empty set.

Case 2

	$5x + 3 = 5(x - 1) + 8$
Use the distributive property.	$5x + 3 = 5x - 5 + 8$
Combine similar terms.	$5x + 3 = 5x + 3$

Subtract 3 from both sides of the equation.	$5x + 3 - 3 = 5x + 3 - 3$
Simplify both sides of the equation.	$5x = 5x$
Subtract $5x$ from both sides of the equation.	$5x - 5x = 5x - 5x$
Simplify both sides of the equation.	$0 = 0$

This is called an *identity* because when you simplified the equation, the left side of the equation equaled the right side. When this happens, any number can be a solution. It doesn't matter what value you use to replace the variable, the left side of the equation will always equal the right side of the equation. Because any real number can be a solution, you have an *infinite* (endless) number of solutions. The mathematical notation for any real number is the capital letter R. You use this R to represent the answer to this equation.

Practice

______**31.** $2x + 5 = 2(x + 5)$

______**32.** $5x + 3x + 2 = 4(2x + 1) - 2$

______**33.** $4x + 3(x + 3) = 7(x + 1) + 2$

______**34.** $3x - 6 = 3(1 + 2x)$

______**35.** $11x - 3 + 4x = 3(5x - 2) + 4$

Applications

Use a variable to write an equation that represents the problem. Then solve the equation to get the answer to the problem.

Example: Your dental insurance company will pay 80% of all fillings after you have met the deductible of $200. For a recent fillings bill, the insurance company paid $176. What was the original dental bill?

Let x = the original dental bill.	$80\%(x - \$200) = \$176$
	$.80(x - 200) = 176$
Use the distributive property.	$.80x - 160 = 176$
Add 160 to both sides of the equation.	$.80x - 160 + 160 = 176 + 160$
Simplify both sides of the equation.	$.80x = 336$
Divide both sides of the equation by .80.	$\frac{.80x}{.80} = \frac{336}{.80}$
Simplify both sides of the equation.	$x = 420$

Example: A trip of 300 miles normally takes you 5 hours when you travel 60 miles an hour. How much faster do you need to drive if you want to make the trip in 4 hours?

Let x = how much faster you need to drive.	$4(60 + x) = 300$
Use the distributive property.	$240 + 4x = 300$
Subtract 240 from both sides of the equation.	$240 - 240 + 4x = 300 - 240$
Simplify both sides of the equation.	$4x = 60$
Divide both sides of the equation by 4.	$\frac{4x}{4} = \frac{60}{4}$
Simplify both sides of the equation.	$x = 15$ miles per hour

36. A corporation pays $\frac{1}{2}$ of the health insurance premium and an additional $200 a month toward other benefits such as dental insurance. Your total benefit package is $450 a month. How much is your health insurance premium each month?

37. A credit card states that your payment will be a minimum of $15 plus 1% of your unpaid balance. Your unpaid balance is $2,365. What is your payment this month?

38. The length of a room is 3 more than twice the width of the room. The perimeter of the room is 66 feet. What are the dimensions of the room? (Let x = the width of the room.)

39. Suppose a roast should be cooked for 45 minutes plus 10 more minutes for every pound the roast weighs. If a roast is properly cooked in 3 hours, how much did it weigh?

40. If you work Christmas Day, your company will pay you double time plus a $50 bonus. If you earn $218 for an 8-hour day, what is your hourly rate of pay?

Skill Building until Next Time

If you have dental work done, what is your deductible? What percent coverage does your insurance provide for preventive care, such as having your teeth cleaned? Is there a difference in coverage when you have a filling, root canal, or crown? If you don't have dental insurance, call an insurance company and inquire about the different types of coverage. What would it cost you to have a cleaning and a root canal? Set up an equation using the information you obtain, and then solve it.

LESSON

Using Formulas to Solve Equations

LESSON SUMMARY

Formulas solve many real-life problems. In this lesson, you will find out what formulas are and how to use them to solve a variety of problems.

Formulas are special equations that show relationships between quantities. For example, you have probably worked with the formula $A=lw$. This formula tells you how to find the area of a rectangle. It tells you to multiply the length times the width to find the area. The formula $D = rt$ tells you to multiply the rate times the time in order to find the distance traveled.

When you substitute the information you know into a formula, you can use that to find the information you don't know. For example, if you travel 55 mph for 3 hours, how far would you travel? Substitute what you know into the equation. Then solve the equation for the variable you don't know.

Substitute what you know into the formula.	$D = rt$
Multiply 55 times 3.	$D = 55 \cdot 3$
You would travel 165 miles.	$D = 165$

What if you wanted to know how long it would take to travel 300 miles if you were traveling at a speed of 60 mph? All you have to do is substitute what you know into the formula, and then solve for the variable you don't know.

$D = rt$

Substitute what you know into the formula. $300 = 60t$

Divide both sides of the equation by 60. $\frac{300}{60} = \frac{60t}{60}$

Simplify both sides of the equation. $5 = t$

It would take you 5 hours to travel 300 miles.

This technique works for any formula even though the formula may be very complex.

Example: Find the interest on a savings account with a balance of $2,400 when the interest rate is 3% for 3 years. Use the formula $I = prt$.

I = interest earned

p = amount of money invested

r = interest rate

t = time invested

$I = prt$

Substitute what you know into the formula. $I = \$2{,}400 \times 3\% \times 3$ years

$I = 2{,}400 \times .03 \times 3$

Simplify the equation. $I = \$216$

Example: How long would you need to invest $3,000 with an interest rate of 3.5% to earn $630?

$I = prt$

Substitute what you know into the formula. $\$630 = \$3{,}000 \times 3.5\% \times t$

$630 = 3{,}000 \times .035t$

Simplify the equation. $630 = 105t$

Divide both sides of the equation by 105. $\frac{630}{105} = \frac{105t}{105}$

Simplify both sides of the equation. $6 = t$

It would take 6 years to earn $630.

Practice

Solve the formulas to find the missing information. Use a calculator to solve these problems.

Use the formula $A = lw$ to solve problems 1 and 2.

A = area of a rectangle

l = length

w = width

1. Find the area of a rectangle when the length is 24.2 cm and the width is 14 cm.

2. Find the length of a rectangle when the width is 12 ft. and the area is 360 ft^2.

Use the formula $I = prt$ to solve problems 3, 4, and 5.

I = interest

p = principal

r = rate

t = time

3. Find the simple interest for a loan of $2,000 with an interest rate of 11.5% for a period of 2 years. (When you enter the interest rate into the formula, enter it as a decimal, not a percent.)

4. How much money will earn $138 interest with a simple interest rate of 11.5% for a period of 2 years?

5. How long will it take to earn $135 interest on $1,500 with an interest rate of 3%?

Use the formula $V = lwh$ to solve problems 6, 7, and 8.

V = volume of a rectangular solid
l = length
w = width
h = height

6. Find the volume of a rectangular prism when the length is 8 ft., the width is 5 ft., and the height is 7 ft.

7. How long would a box need to be if $V = 270$ cm^3, $w = 6$ cm, and $h = 3$ cm?

8. What is the height of a box if the $V = 858$ ft.^{3}, $l = 13$ ft., and $w = 11$ ft?

Clark's Rule is commonly used to calculate the children's dosage of a medication. Clark's Rule is: $C = A \cdot \frac{w}{150}$ where

C = child's dose
A = adult's dose
w = weight of child in pounds

Use Clark's Rule to solve problems 9, 10, and 11.

9. If the adult's dosage is 60 cc, find the child's dosage if the child weighs 50 lbs.

10. If the child weighs 90 lbs. and the adult dosage is 70 cc, what is the correct dosage to give a child?

11. How much would a child need to weigh to receive the adult dosage if the adult dosage is 80 cc?

The force exerted by an object depends on its mass and the amount by which it is accelerated. Use the formula $F = ma$ to solve problems 12, 13, and 14.

F = force in Newtons
m = mass in kilograms
a = acceleration in meters per second squared $(\frac{m}{s}{}^2)$

12. How much force is exerted by a 5 kg cement block accelerated by gravity $(9.8\frac{m}{s^2})$?

13. How many kilograms of mass are needed to exert 100 Newtons of force under the acceleration due to gravity $(9.8\frac{m}{s^2})$?

14. If a 50-kg mass exerts a force of 10,000 Newtons, how fast is it accelerating?

Suppose that the monthly expenses of a factory are given by: $E = pm + R$. Use this formula to solve problems 15, 16, and 17.

E = monthly expenses
p = number of items made each month
m = cost of the raw materials to make one item
R = monthly rent.

15. Suppose the rent on a factory is $15,000 per month and it makes 5000 items each month. If the raw materials to produce one item cost $12, what are the total montly expenses?

16. Suppose the total monthly expenses of a factory are $35,000. If the monthly rent is $15,000 and the factory produced 100,000 items, how much do the raw materials to produce one item cost?

17. Suppose a factory will produce 500 items a month and the raw materials to produce one item cost $30. If the total monthly expenses must be kept below $25,000, what is the maximum monthly rent that can be afforded?

The Occupational Safety and Health Administration (OSHA) established permissible sound exposures in the workplace. This is modeled by the equation: $S = -2.817H + 108.9$. Use this formula to solve problems 18, 19, and 20.

S = maximum permissible sound level in decibels
H = number of hours of exposure

18. What is the maximum permissible sound level in decibels for an 8-hour work day?

19. What is the maximum permissible sound level in decibels for a 3-hour exposure?

20. How long can you be exposed to receive a maximum sound level of 105 decibels?

The formula $F = \frac{9}{5}C + 32$ can be used to make temperature conversions with the Celsius and Fahrenheit temperature scales. Use this formula for problems 21, 22, and 23.

F = Fahrenheit temperature
C = Celsius temperature

21. Convert 100° Fahrenheit to the Celsius temperature scale.

22. The boiling point in the Celsius scale is 100°. What is the boiling point for the Fahrenheit scale?

23. If ice freezes at 32° F, what will the temperature be on the Celsius scale?

The area of a trapezoid depends on the lengths of its two ends and the distance between them. Use the formula $A = \frac{1}{2}(b_1 + b_2)h$ to solve problems 24, 25, 26, and 27.

A = area of a trapezoid
b_1 = width at the top
b_2 = width at the bottom
h = height

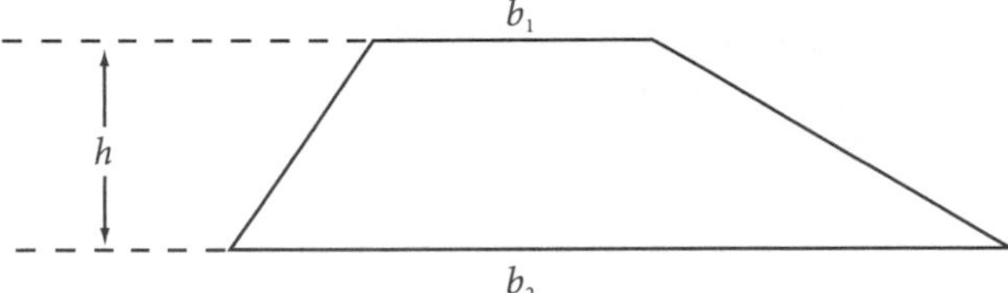

24. What is the area of a trapezoid that is 5 feet tall, 4 feet wide at the top, and 10 feet wide at the bottom?

25. How tall is a trapezoid that is 4 inches across the top, 12 inches across the bottom, and has 48 square inches of area?

26. A 7-foot tall trapezoid has 42 square feet of area. If it is 9 feet wide at the bottom, how far across is it at the top?

27. Suppose a trapezoid is 10 cm tall and has 80 cm^2 of area. If the distance across the top is the same as the distance across the bottom, what is this distance?

The annual rate of depreciation of a car can be determined using the formula: $D = \frac{C-S}{N}$. Use the formula to solve problems 28, 29, and 30.

D = amount to be depreciated each year
C = original cost
S = scrap value
N = estimated life

28. Find the annual depreciation of a GMC pickup whose original cost was $29,999. The estimated life of the pickup is 20 years and the scrap value will be $600.

29. How much will the GMC pickup in problem 28 be worth after 8 years?

30. Find the annual depreciation of a Dodge Neon with an original cost of $9,500. The estimated life of the car is 10 years with a scrap value of $300.

Skill Building until Next Time

Several businesses have electronic signs that display the current temperature. The next time you drive through the business section of a city, pay attention to the current temperature. Use the formula $F = \frac{9}{5}C + 32$ to convert the current temperature from the scale on the sign to the other scale. If the sign gives the temperature in both Celsius and Fahrenheit, use the formula to verify that the two temperatures are the same.

LESSON

Graphing Linear Equations

LESSON SUMMARY

You have solved equations using algebra. Most of the equations you solved had one variable and one solution. This lesson teaches you how to solve equations with two variables by using a graph. These equations have many solutions. You many want to get some graph paper to use during this lesson.

▶ What Is a Graph?

A *graph* is a picture. When you graph an equation, you are creating a picture of the answers. To graph equations with two variables, you use a coordinate plane. Two intersecting lines that meet at right angles form a *coordinate plane*.

The two intersecting lines that form the coordinate plane intersect at a point called the *origin*. The origin is your starting point. The horizontal line is called the *x-axis*. When you move to the right of the origin on the x-axis, the numbers are positive. When you move to the left of the origin on the x-axis, the numbers are negative. The vertical axis is called the *y-axis*. When you move above the origin on the y-axis, the numbers are positive. When you move below the origin, the numbers are negative.

The x and y axes divide the plane into four equal parts. These parts are called *quadrants* and are named by a number. The quadrant in the upper right hand corner is quadrant 1. To label the other quadrants, go counter-clockwise.

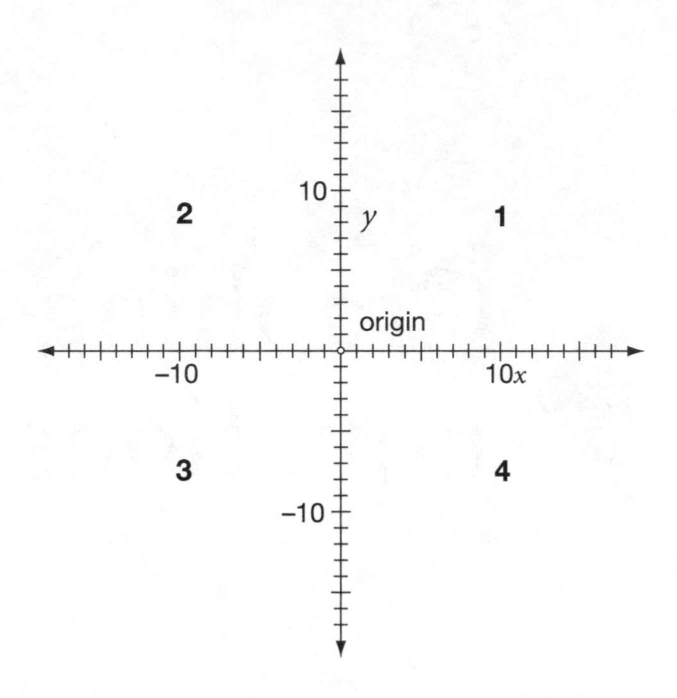

▶ Plotting Points on a Graph

You can plot points on a coordinate plane. Every point has two coordinates: an x coordinate and a y coordinate. These coordinates are written as an *ordered pair*. It is called an ordered pair because it is a pair of numbers with a special order. The pair of numbers is enclosed in parentheses with the x coordinate first and the y coordinate second. For example, the ordered pair (2,3) has an x coordinate of 2 and a y coordinate of 3. To plot this point, start at the origin and move 2 units in the positive direction on the x-axis. From there, move up 3 units because the y coordinate is a positive 3.

To plot the ordered pair (4,2), you start at the origin and move 4 units in the positive direction on the x-axis. From there, move 2 units up because 2 is positive. To graph the point (–5,3), you start at the origin and move 5 units in the negative direction on the x-axis. From there, move up 3 units. To graph (–2,–4), you start at the origin and move 2 units in the negative direction on the x-axis. From that point, move down 4 units because the 4 is negative.

Example: Look at the points on the coordinate plane below. Each letter names an ordered pair. The ordered pairs are listed below the coordinate plane.

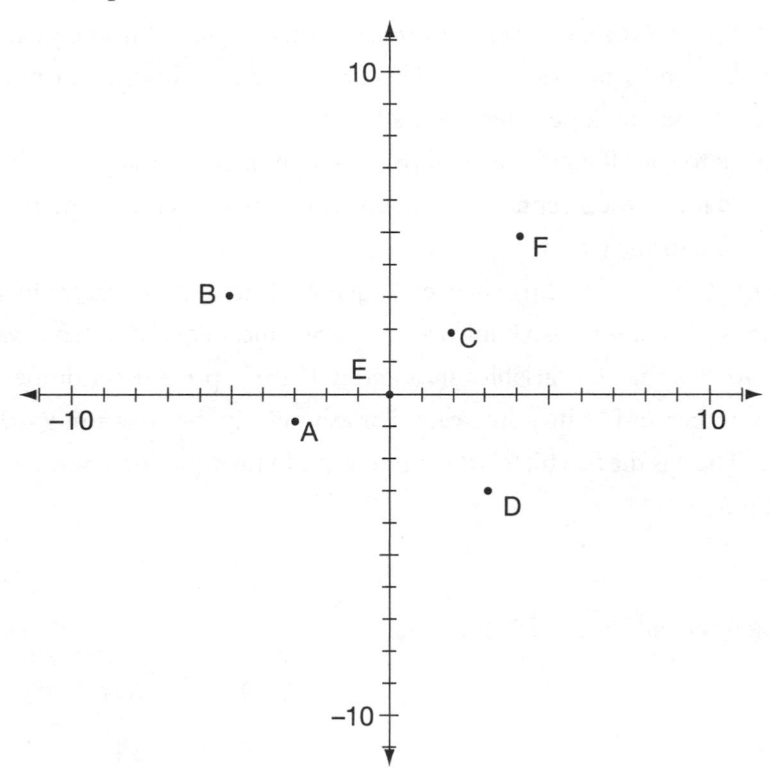

A (–3,–1)
B (–5,3)
C (2,2)
D (3,–3)
E (0,0)
F (4,5)

Practice

Using graph paper, plot the following points. You can plot all of these points on the same coordinate plane.

1. (8,2)

2. (8,–2)

3. (0,4)

4. (5,0)

5. (–5,–5)

6. (–1,5)

7. (6,–3)

8. (–6,–2)

9. (2,9)

10. (–7,3)

▶ Using the Slope and *Y*-Intercept

Did you find plotting points easy? If so, you are ready to move on and graph linear equations, but first, you need to know what the terms *slope* and *y-intercept* mean. There are several methods that can be used to graph linear equations; however, you will use the slope-intercept method here.

What does slope mean to you? If you are a skier, you might think of a ski slope. The slope of a line has a similar meaning. The *slope* of a line is the steepness of a line. What is the y-intercept? Intercept means to cut, so the y-intercept is where the line cuts the y-axis.

To graph a linear equation, you will first change its equation into slope-intercept form. The slope-intercept form of a linear equation is $y = mx + b$; also known as $y =$ *form*. Linear equations have two variables. For example, in the equation $y = mx + b$, the two variables are x and y. The m represents a number and is the slope of the line. The b represents a number and is the y-intercept. For example, in the equation $y = 2x + 3$, the number 2 is the m, which is the slope. The 3 is the b, which is the y-intercept. In the equation $y = -3x + 5$, the m or slope is -3, and the b or y-intercept is 5.

Practice

Find the slope (m) and y-intercept (b) of each equation.

_______**11.** $y = 3x + 9$

_______**12.** $y = 5x - 6$

_______**13.** $y = -5x + 16$

_______**14.** $y = -2.3x - 7.5$

_______**15.** $y = \frac{3}{4}x + 5$

_______**16.** $y = \frac{x}{3} + 8$ (Hint: The number in front of the x is 1.)

Getting the Right Form

What if the equation is not in slope-intercept form? Simple! All you need to do is change the equation to slope-intercept form. How? Slope-intercept form is $y =$ *form*, so your strategy is to get the y on a side by itself.

An equation needs to be in slope-intercept form or $y =$ *form* ($y = mx + b$) before you can graph the equation with a pencil and graph paper. Also, if you were to use a graphing calculator to graph a linear equation, the equation needs to be in $y =$ *form* before it can be entered into the calculator.

Example: $2x + y = 5$

Subtract $2x$ from both sides of the equation.	$2x - 2x + y = 5 - 2x$
Simplify.	$y = 5 - 2x$
Rearrange the equation so the x term is first.	$y = -2x + 5$

A mathematical rule called the *commutative property* lets you change the order of numbers or terms when you add or multiply. You want the previous equation in the form $y = mx + b$, so the order of the 5 and the $-2x$ needed to be changed after getting the y on a side by itself. When you move a term, be sure to take the sign of the term with it. For example, the 5 was a positive number. It remains a positive number when you move it.

Example: $2x + 3y = 9$

Subtract $2x$ from both sides of the equation.	$2x - 2x + 3y = 9 - 2x$
Simplify.	$3y = 9 - 2x$
Use the commutative property.	$3y = -2x + 9$
Divide both sides by 3.	$y = -\frac{2x}{3} + \frac{9}{3}$
Simplify both sides of the equation.	$y = -\frac{2}{3}x + 3$

Tip

Because $\frac{2x}{3}$ and $\frac{2}{3}x$ name the same number, you may see it written either way.

Example: $-3x + 2y = 10$

Add $3x$ to both sides of the equation.	$-3x + 3x + 2y = 10 + 3x$
Simplify.	$2y = 10 + 3x$
Use the commutative property.	$2y = 3x + 10$
Divide both sides of the equation by 2.	$\frac{2y}{2} = \frac{3x}{2} + \frac{10}{2}$
Simplify both sides of the equation.	$y = \frac{3}{2}x + 5$

Practice

Change the equations into slope-intercept form. Then state the slope (m) and y-intercept (b).

______**17.** $2x + y = -4$

______**18.** $3x + y = 6$

______**19.** $-3x + y = 8$

______**20.** $5x + 5y = 15$

______**21.** $-20x + 10y = 50$

______**22.** $-3x + 4y = 12$

______**23.** $8x - 2y = 6$

______**24.** $-3x + 6y = 12$

______**25.** $-6x - 3y = 5$

______**26.** $-4x + 5y = 0$

▶ Graphing Linear Equations Using the Slope and *Y*-Intercept

The graph of a linear equation is a line, which means it goes on forever in both directions. A graph is a picture of all the answers to the equation, so there is an infinite (endless) number of solutions. Every point on that line is a solution.

Tip

Look at all coordinate planes carefully to see what scale each is drawn to—some may have increments of one, while others may have increments of one-half or ten.

You know that slope means the steepness of a line. In the equation $y = 2x + 3$, the slope of the line is 2. What does it mean when you have a slope of 2? Slope is defined as the rise of the line over the run of the line. If the slope is 2, this means $\frac{2}{1}$, so the rise is 2 and the run is 1.

If the slope of a line is $\frac{2}{3}$, the rise is 2 and the run is 3. What do rise and run mean? *Rise* is the vertical change, and *run* is the horizontal change. To graph a line passing through the origin with a slope of $\frac{2}{3}$, start at the origin. The rise is 2, so from the origin go up 2 and to the right 3. Then draw a line from the origin to the endpoint. The line you have drawn has a slope of $\frac{2}{3}$.

Now draw a line with a slope of $-\frac{3}{4}$. Start at the origin. Go down 3 units because you have a negative slope. Then go right 4 units. Finally, draw a line from the origin to the endpoint. These two lines appear on the same graph below.

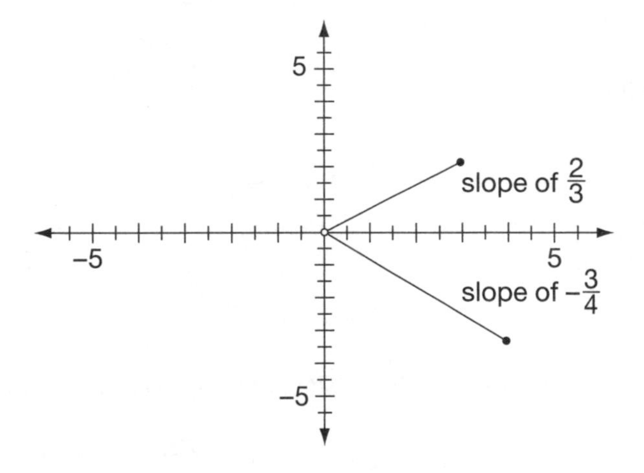

Tip

A positive slope will always rise from left to right. A negative slope will always fall from left to right.

To graph an equation like $y = x + 1$, you can use the slope and y-intercept. The first step is to determine the slope. The slope is the number in front of x, which means, in this case, it is 1. What is the y-intercept? It is also 1. To graph the equation, your starting point will be the y-intercept, which is 1. From the y-intercept, use the slope, which is also 1, or $\frac{1}{1}$. The slope tells you to go up 1 and to the right 1. A line is drawn from the origin (the y-intercept) to the endpoint (1,2). You can extend this line and draw arrows on each end to show that the line extends infinitely.

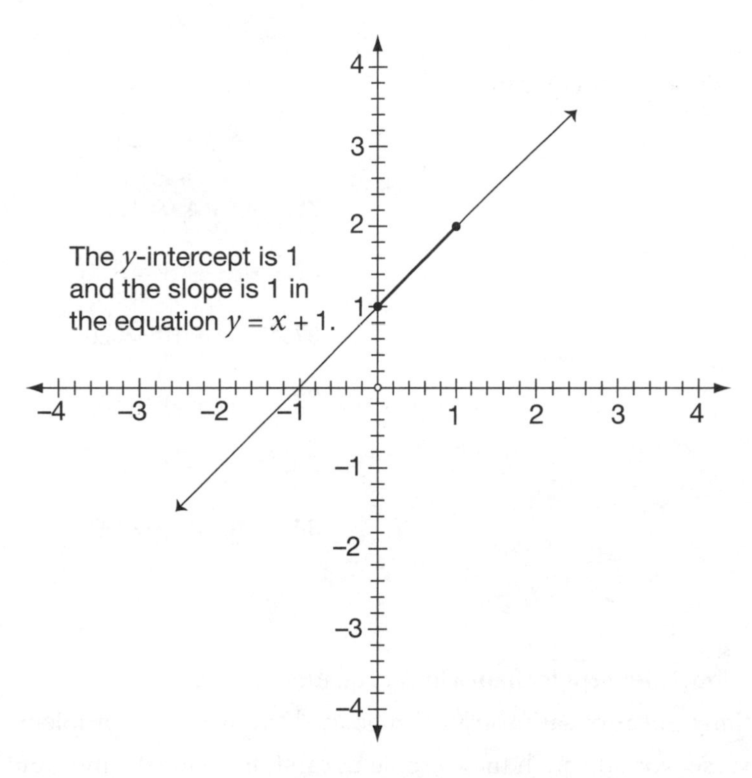

Example: Graph the equation $y = -\frac{2}{3}x + 2$. Start with the y-intercept, which is a positive 2. From there, go down 2 because the slope is negative and to the right 3. Draw a line to connect the intercept and the endpoint.

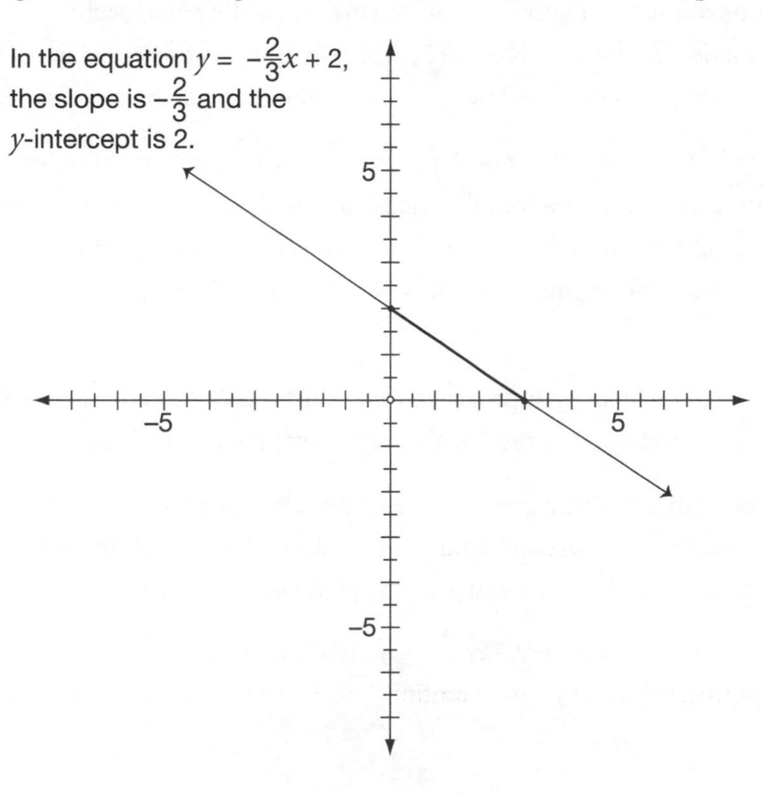

Practice

Use graph paper to graph these linear equations.

27. $y = x + 4$

28. $y = 2x + 3$

29. $y = 3x - 2$

30. $y = -2x + 5$

31. $y = \frac{3}{2}x + 1$

32. $y = -\frac{2}{3}x + 5$

33. $-\frac{1}{4}x + y = 2$

34. $-8 + 4y = -12$

35. $4x + 4y = 12$

36. $-8x + 4y = 8$

37. $-5x + 10y = 20$

38. $-4x + 3y = 12$

39. $x + y - 3 = 0$

40. $-3x + 4y = 12$

Applications

Writing equations in the slope-intercept form of a linear equation, $y = mx + b$, can be useful in solving word problems. Additionally, graphing linear equations can also be used to solve word problems. Word problems are an important part of algebra, so work through these problems carefully to get the most out of them.

Example: If an airplane maintains a landing approach of a constant rate of descent of 50 ft. for every 500 ft. horizontally, what is the slope of the line that represents the plane's landing approach?
Solution: The rise of the landing approach is down 50 which would be represented by $a - 50$. The run of the landing approach is 500. The slope is represented by the rise over the run, which is $-\frac{50}{500} = -\frac{1}{10}$.

Example: You are charged a flat fee of \$5 a month plus \$.11 per kilowatt-hour of power used. Write an equation that would calculate your power bill for a month. Then state the slope and y-intercept of your equation.
Solution: You will be charged \$.11 for each kilowatt-hour of power you use, so let x = the number of kilowatt hours. Let y = the monthly power bill. Your equation will be $y = .11x + 5$. The slope of the equation is .11, and the y-intercept is 5.

41. An airplane maintains a landing approach of a constant rate of descent of 70 ft. for every 700 ft. traveled horizontally. What is the slope of the line that represents the plane's landing approach?

42. You are a sales clerk in a clothing store. You receive a salary of \$320 a week plus a 5% commission on all sales. Write an equation to represent your weekly salary. (Let y equal the weekly salary and x equal the amount of sales.) What is the slope and y-intercept of your equation?

43. You are renting a car from Economy Rent-A-Car. It will cost you \$25 a day plus \$.10 a mile per day. Write an equation to represent the daily cost of renting the car. What is the slope and y-intercept of your equation?

44. A phone plan charges \$5 a month plus \$.05 for every minute used each month. Write an equation to represent the total monthly cost of the phone plan. Graph the linear equation.

45. When you travel at a speed of 60 mph., write an equation that represents how far you will travel in x hours. Then graph the linear equation. (When you set your scale on your graph, let each unit represent 1 hour on the x scale and 10 miles on the y scale. You can use different scales on the x and y axes so that your graph is a reasonable size.)

Skill Building until Next Time

What are some real-world examples of slope that you can think of besides a ski slope? Pay attention to your surroundings today and observe examples of slope being used in the real world. Perhaps it's the angle of a stairway, or the way a branch is leaning off the trunk of a tree. Try to find at least five examples of slope as you move through the day.

LESSON

Solving Inequalities

LESSON SUMMARY

An equation is two equal expressions connected with an equal sign. Is it possible to have two expressions that are not equal? Yes! Two expressions that are not equal are called inequalities, and like equations, you can solve inequalities. In this lesson, you will find that the methods you use to solve inequalities are similar to those you use to solve equations.

▶ What Is an Inequality?

An *inequality* is two numbers or expressions that are connected with an inequality symbol. The inequality symbols are $<$ (less than), $>$ (greater than), $\leq$ (less than or equal to), $\geq$ (greater than or equal to), and $\neq$ (not equal to). Here are some examples of inequalities:

$2 < 5$ (two is less than five)
$9 > 3$ (nine is greater than three)
$4 \leq 4$ (four is less than or equal to four)
$2x + 5 \neq 11$ ($2x$ added to five is not equal to eleven)

▶ Solving Inequalities

You can solve inequalities with variables just like you can solve equations with variables. Use what you already know about solving equations to solve inequalities. Like equations, you can add, subtract, multiply, or divide both sides of an inequality with the same number. In other words, what you do to one side of an inequality, you must do to the other side.

Example: $2x + 3 < 1$

Subtract 3 from both sides of the inequality.	$2x + 3 - 3 < 1 - 3$
Simplify both sides of the inequality.	$2x < -2$
Divide both sides of the inequality by 2.	$\frac{2x}{2} < \frac{-2}{2}$
Simplify both sides of the inequality.	$x < -1$

The answer for this example is the inequality, $x < -1$. There are an infinite (endless) number of solutions because every number less than –1 is an answer. In this problem, the number –1 is not an answer because the inequality states that your answers must be numbers less than –1.

Tip

The answer to an inequality will always be an inequality. Because the answer is an inequality, you will have an infinite number of solutions.

Did you notice the similarity between solving equations and solving inequalities? You can see that the previous example was solved using the same steps you would use if you were solving an equation.

However, there are some differences between solving equations and solving inequalities. Notice what happens when you multiply or divide an inequality by a *negative* number.

$$2 < 5$$

$$-2 \cdot 2 < 5 \cdot -2$$

$$-4 < -10$$

However, –4 is not less than –10. So $-4 < -10$ is a false statement. To correct it, you would have to rewrite it as $-4 > -10$.

Tip

Whenever you multiply or divide an inequality by a negative number, you need to reverse the inequality symbol.

You can solve inequalities using the same methods that you use to solve equations with these exceptions:

1. The answer to an inequality will always be an inequality.
2. When you multiply or divide an inequality by a negative number, you must reverse the inequality symbol.

Practice

Solve the inequalities. Be sure to write out your steps.

______ **1.** $x + 3 < 10$

______ **2.** $x - 10 \geq -3$

______ **3.** $2x + 5 < 7$

______ **4.** $5x + 3 \leq -4$

______ **5.** $-3x \leq 9$

______ **6.** $4x \geq -32$

______ **7.** $\frac{x}{3} < 5$

______ **8.** $\frac{x}{-4} < 2$

______ **9.** $\frac{x}{-3} > 5 - 9$

______ **10.** $5x + 1 \leq 11$

______ **11.** $-3x + 6 < 24$

______ **12.** $6x + 5 > -11$

______ **13.** $x + 5 \leq 4x - 4$

______ **14.** $\frac{x}{4} - 2 \leq 10$

______ **15.** $-3x > 2x + 10$

______ **16.** $4x + 2 \geq 6x - 1$

______ **17.** $10x > 0$

______ **18.** $3x + 2(x - 1) \geq x + 7$

______ **19.** $\frac{3x}{4} < -6$

______ **20.** $\frac{x}{-5} + 2 > 2$

______ **21.** $2.2x - 2 \geq 2.4$

______ **22.** $5x + 8 \leq 2x - 1$

______ **23.** $2(x + 4) > 10$

______ **24.** $2(x + 5) > 3(x - 2)$

______ **25.** $-2x - 8 > 3(x + 4)$

______ **26.** $\frac{x}{2} + 9 < x + 11$

► Checking Your Answers

There is a way to check your answer to an inequality. Just follow the steps shown in the following example.

Example: $x + 4 < 10$

Subtract 4 from both sides of the inequality.	$x + 4 - 4 < 10 - 4$
Simplify both sides of the inequality.	$x < 6$

The answer is the inequality, $x < 6$. This means that any number less than 6 is an answer. To check your answer, pick a number less than 6 and substitute that number into the *original* problem. Let's use the number 2 to check the answer.

Check: $x + 4 < 10$

Substitute 2 in place of the variable.	$2 + 4 < 10$
Simplify.	$6 < 10$

Yes, 6 is less than 10. When your result is a true statement, you know you have worked the problem correctly. Try your hand at checking some inequalities—how about practice problems 1–10 in this lesson?

Applications

Write an inequality and then solve each of the following word problems. To get started, read through the two example problems.

Example: You are taking your friend to the fair. You have \$40 to spend. What is the maximum amount of money you can spend on each of you?

Let x represent what you can spend. Since there are two of you, this would be represented by $2x$. You can spend less than or equal to \$40, so your inequality would be:

	$2x \leq \$40$
Divide both sides of the inequality by 2.	$\frac{2x}{2} \leq \frac{40}{2}$
Simplify.	$x \leq \$20$

Example: If the admittance price for a movie is \$6 and you have \$10 to spend, how much can you spend on concessions?

Let x equal the amount of money you can spend on concessions.

	$x + 6 \leq 10$
Subtract 6 from both sides of the inequality.	$x + 6 - 6 \leq 10 - 6$
Simplify both sides of the inequality.	$x \leq \$4$

Practice

27. You are treating a friend to a movie. You will buy 2 tickets and spend \$8 on concessions. If you don't want to spend more than \$20, how much can you spend on the tickets?

28. You are pricing lawn furniture and plan to buy 4 chairs. You don't want to spend more than \$120. What is the most you can spend on one chair?

29. You are going to a restaurant for lunch. You have \$15 to spend. Your beverage is \$2.50, and you will leave a \$2 tip. How much can you spend on the entrée?

30. There are only 2 hours left in the day and Amy still has 5 people left to interview. What is the maximum amount of time she can spend on each interview?

Skill Building until Next Time

When you comparison shop, you find that the price of an item may vary from store to store. If the wholesale price of an item is \$50, the store owner needs to sell the item for more than \$50 in order to make a profit. Why would the retail prices vary from store to store if all the stores pay the same wholesale price?

LESSON

10 Graphing Inequalities

LESSON SUMMARY

In this lesson, you will graph inequalities with one and two variables. To graph an equation or inequality with one variable, you use a number line. To graph equations or inequalities with two variables, you use a coordinate plane.

What Is a Number Line?

A row of numbers represented on a line is called a *number line.* The starting point is 0. If you move to the right of 0 on the number line, the numbers are positive and become increasingly larger. If you move to the left of 0, the numbers are negative and get increasingly smaller. Here is an example of a number line.

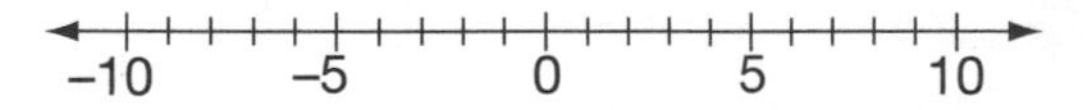

Graphing Points and Lines on a Number Line

To graph $x = 3$, you put a point on the number line at 3. To plot the point –5, you put a point on –5.

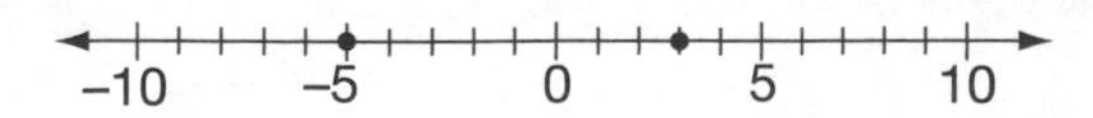

To graph the inequality $x > 5$, put an open circle on 5, then draw a line and arrow to the right. The open circle shows that 5 is not an answer. The line shows that *all* the points greater than 5 are answers. The arrow shows that the answers are endless (infinite).

To graph $x \leq -2$, put a closed circle on –2 and a line and arrow going to the left of –2. The closed circle shows that –2 is an answer. The line shows that every number less than –2 is also an answer, and the arrow shows that the answers are endless.

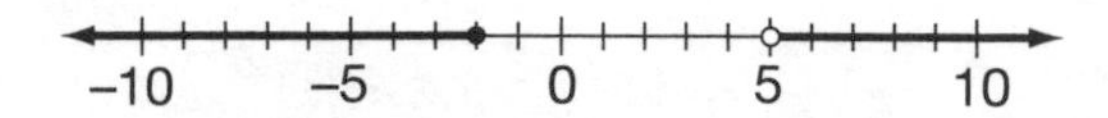

Tip

An open circle on a number line tells you the endpoint *is not* included. A closed circle tells you the endpoint is included and that the number *is* an answer. The arrow at the end of the line indicates that the line goes on forever and the answers are infinite.

Practice

Graph the following equations and inequalities on a number line.

1. $x = -3$
2. $x < -3$
3. $x > 1\frac{1}{2}$
4. $x \leq 4$
5. $x \geq 0$

▶ Graphing Linear Inequalities

A *linear inequality* has two variables just like a linear equation. The inequality $2x + y < 1$ is a linear inequality with two variables. You can draw on what you already know to graph linear inequalities. A linear equation graphs into a line. A linear inequality has two parts: a line and a shaded area.

When you graphed linear equations, your first step was to put the equation into $y =$ *form.* Do the same with the linear inequality. Do you remember the mathematical rule called the *commutative property* that you learned in Lesson 8? It lets you change the order of numbers or terms when you add or multiply. When you move a term, be sure to take the sign of the term with it.

Tip

If the inequality symbol is < or >, the boundary line will be dotted. If the inequality symbol is ≤ or ≥, the boundary line will be solid.

Example: $2x + y > 1$

Subtract $2x$ from both sides of the inequality.	$2x - 2x + y > 1 - 2x$
Simplify.	$y > 1 - 2x$
Use the commutative property.	$y > -2x + 1$

The inequality $y > -2x + 1$ tells you that the slope is -2 and the y-intercept is 1. If the inequality has a $>$ or $<$ symbol, then the line will be dotted. If the inequality symbol is $\leq$ or $\geq$, then the line will be solid. Generally, if the inequality symbol is $>$ or $\geq$, you shade above the line. If the inequality symbol is $<$ or $\leq$, you shade below the line.

To graph $y > -2x + 1$, start with the y-intercept, which is 1. The slope is -2, which means $\frac{-2}{1}$, so from the y-intercept of 1, go down 2 because the slope is negative and to the right 1. When you connect the starting point and the ending point, you will have the boundary line of your shaded area. You can extend this line as far as you'd like in either direction since it is endless. The boundary line will be dotted because the inequality symbol is $>$. If the symbol had been $\geq$, then the line would be solid, not dotted.

When you graph, always check your graph to make certain the direction of the line is correct. If the slope is positive, the line should rise from left to right. If the slope is negative, the line should fall from left to right.

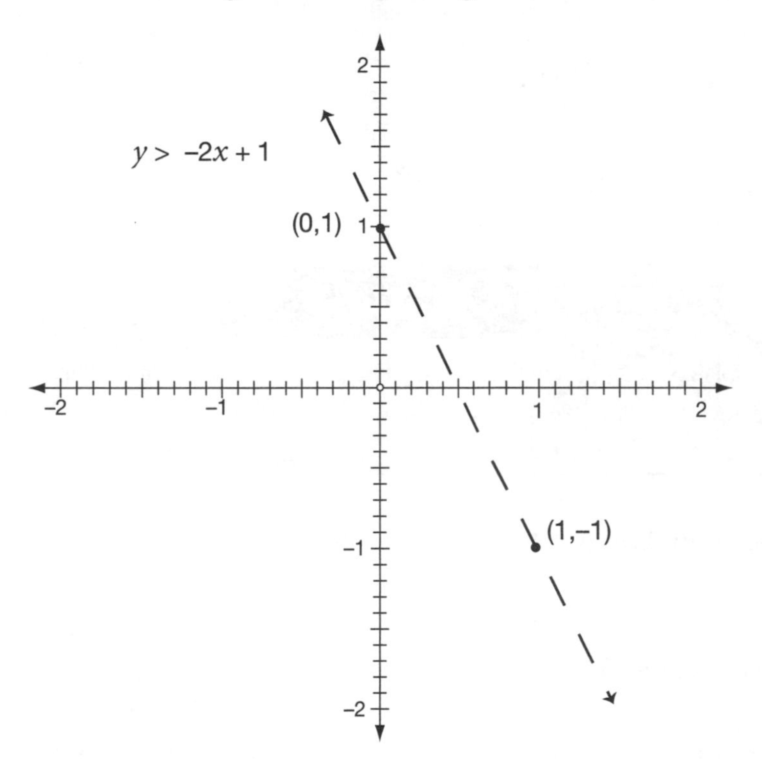

Now that you have the boundary line, will you shade above or below the line? The inequality symbol is >, so shade above the line.

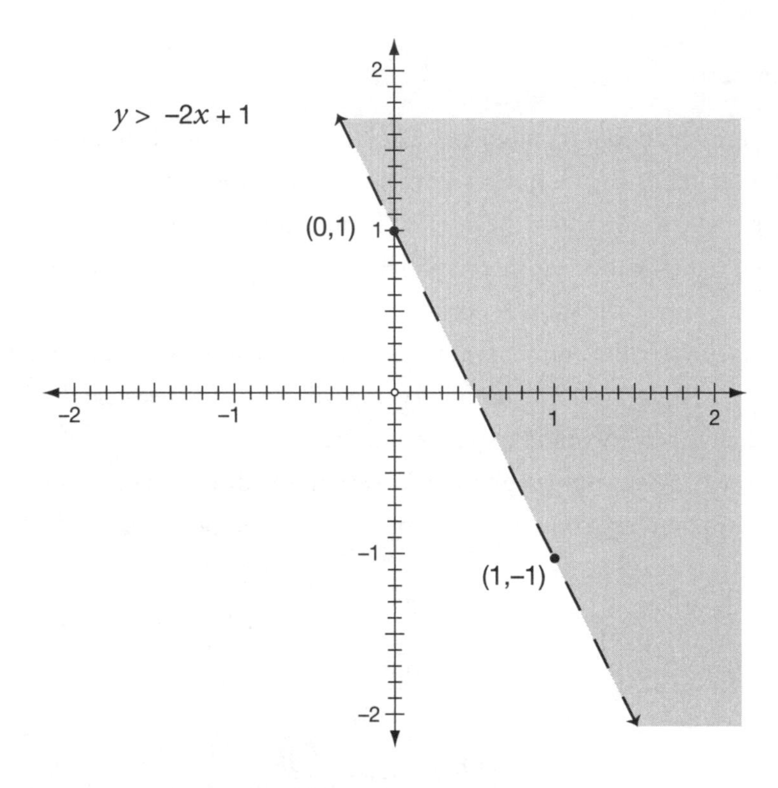

Tip

If the inequality symbol is > or ≥, you will generally shade above the boundary line. If the inequality symbol is < or ≤, you will shade below the boundary line.

Example: Graph the inequality $3x + 2y \leq 4$. (Hint: Remember that the first step is to change the equation to $y =$ *form*.)

	$3x + 2y \leq 4$
Subtract $3x$ from both sides of the inequality.	$3x - 3x + 2y \leq 4 - 3x$
Simplify.	$2y \leq 4 - 3x$
Use the commutative property.	$2y \leq -3x + 4$
Divide both sides of the inequality by 2.	$\frac{2y}{2} \leq -\frac{3x}{2} + \frac{4}{2}$
Simplify both sides of the inequality.	$y \leq -\frac{3}{2}x + 2$

The y-intercept of the inequality is 2 and the slope is $-\frac{3}{2}$. Start with the y-intercept, which is 2. From that point, go down 3 because the slope is negative and to the right 2. Then connect the starting point and the ending point. Will the boundary line be dotted or solid? It will be solid because the inequality symbol is ≤. Will you shade above or below the boundary line? You will shade below the boundary line because the inequality symbol is ≤.

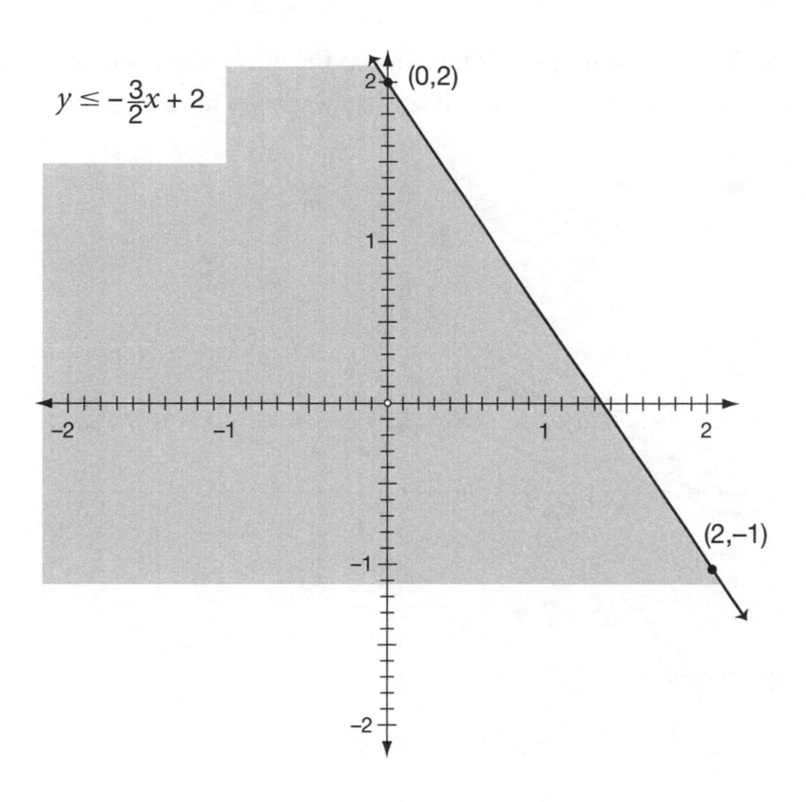

Practice

Use graph paper to graph the following inequalities.

6. $x + y > 1$

7. $-2x + y < 2$

8. $6x + 2y \le 8$

9. $-4x + 2y \ge 6$

10. $x - y > 5$

11. $3x - y < 2$

12. $x - 3y \ge 9$

13. $-.5x + y > 1$

14. $-.6x + .3y < .12$

15. $y - 3 > 2x + 1$

16. $-\frac{2}{3}x + y \le 2$

17. $3x - 2y \le 5$

► Special Cases of Inequalities

There are two special cases of inequalities. One has a vertical boundary line and the other has a horizontal boundary line. For example, the inequality $x > 2$ will have a vertical boundary line, and the inequality $y > 2$ will have a horizontal boundary line. The inequality $y > 2$ is the same as the inequality $y > 0x + 2$. It has a slope of 0 and a y-intercept of 2. When the slope is 0, the boundary line will always be a horizontal line.

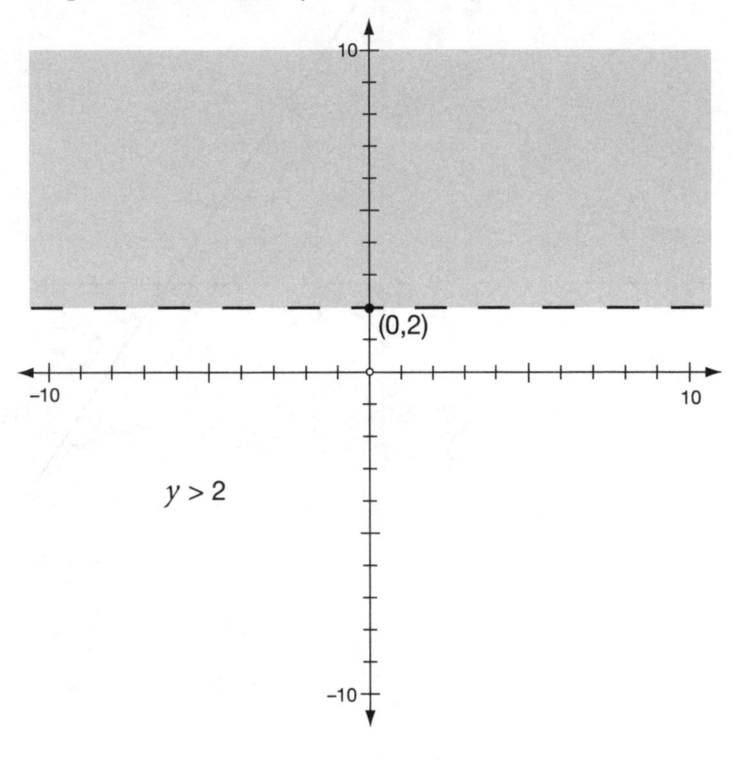

The inequality $x > 2$ cannot be written in $y =$ *form* because it does not have a slope or a y-intercept. It will always be a vertical line. It will be a dotted vertical line passing through where $x = 2$.

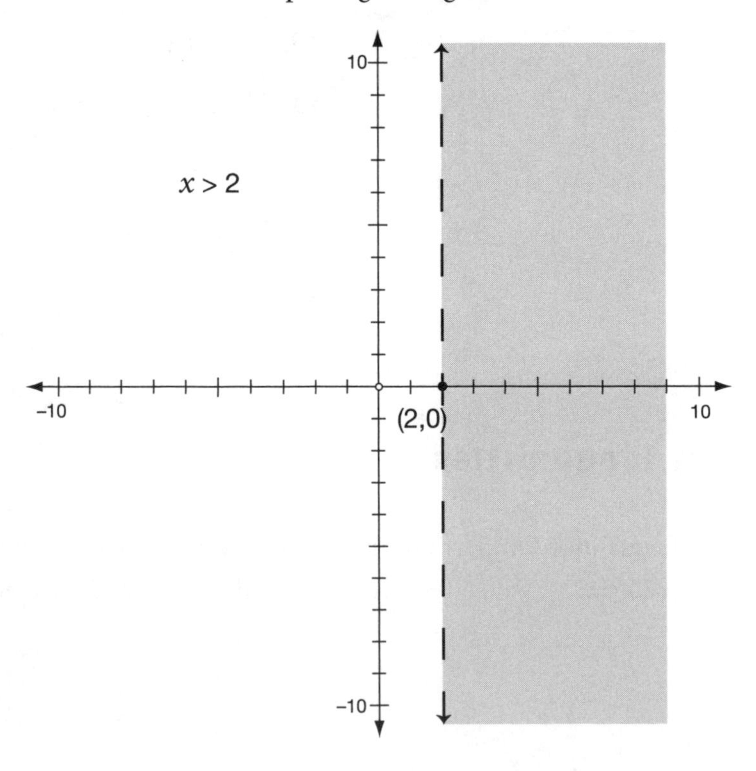

Tip

A horizontal line has a slope of 0. A vertical line has no slope.

Practice

Graph the special cases of inequalities.

18. $x > 5$

19. $y < -4$

20. $x \leq -2$

21. $y \geq 0$

Applications

Write an inequality and then solve each word problem on the next page. To get started, read through the following example problem.

Example: You ordinarily get paid \$5 an hour at your part-time job. You can earn \$10 an hour if you work on a holiday. How many hours would you have to work on Christmas to earn at least \$100? Let x = the number of regular hours worked and let y = the number of hours worked on a holiday. Write an inequality to solve the problem and solve it graphically.

	$5x + 10y \geq \$100$
Subtract $5x$ from both sides of the inequality.	$5x - 5x + 10y \geq \$100 - 5x$
Simplify.	$10y \geq 100 - 5x$
Use the commutative property.	$10y \geq -5x + 100$
Divide both sides of the inequality by 10.	$\frac{10y}{10} \geq -\frac{5x}{10} + \frac{100}{10}$
Simplify both sides of the inequality.	$y \geq -\frac{1}{2}x + 10$

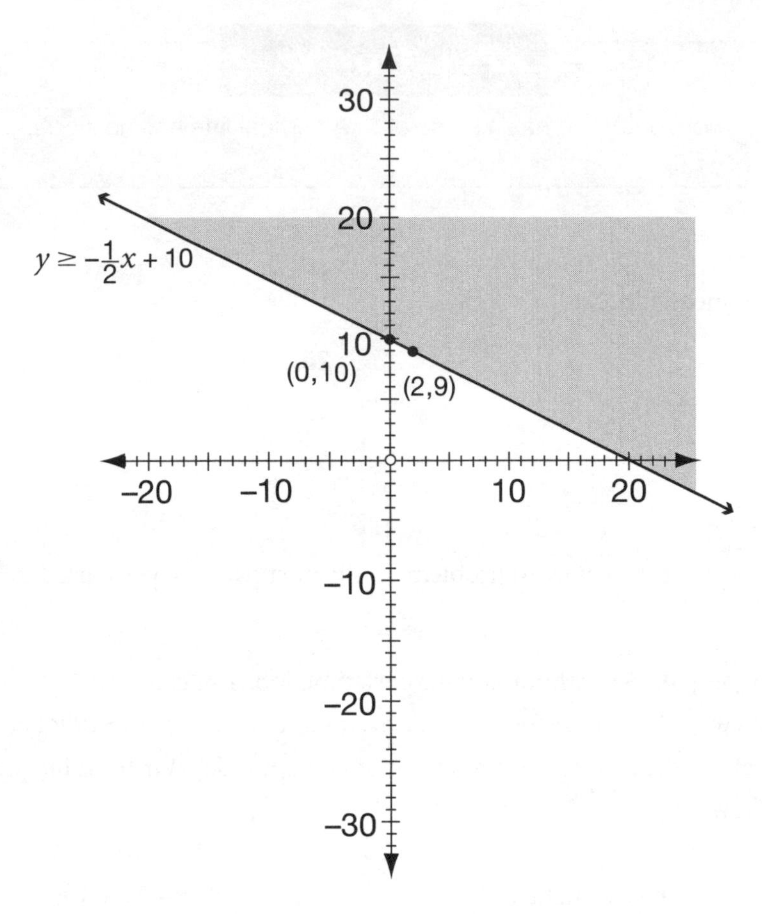

22. You want to spend no more than \$100 on CDs and tapes. The CDs will cost \$15 each and the tapes will cost \$10 each. Write an inequality to solve the problem, and then graph the inequality. (Let x equal the number of CDs and y equal the number of tapes.)

23. You have two part-time jobs. You get \$6 an hour as a delivery person for the Pizza Barn. You also get paid \$5 an hour for doing yard work. How many hours will you need to work at both jobs to earn at least \$100? Write an inequality to solve the problem. Then solve the problem graphically. (Let x equal the number of hours worked at Pizza Barn and y equal the number of hours doing yard work.)

24. There are two cookie recipies. One uses 2 cups of flour in each batch and the other uses $1\frac{1}{2}$ cups of flour in each batch. If you have only six cups of flour, how many batches of each can you make? Write an inequality to solve the problem, and then graph the inequality.

25. A basketball player wants to score at least 20 points in a game. He gets 2 points for a field goal and 1 point for each free throw. Write an inequality to solve the problem, and then solve the problem graphically. (Let x equal the number of field goals and y equal the number of free throws.)

Skill Building until Next Time

When you graph an inequality, you have a picture of all possible answers to a problem. Can you see that it would be impossible to list them all? Think how valuable graphs are to the business world. If the shaded area on a graph represents all the possible prices a company can charge for a product and make a profit, executives can easily determine what to charge for a product to achieve the desired profit. Consider how airline ticket prices can vary on the same flight. Why do you think the airlines do this?

LESSON

11 Graphing Systems of Linear Equations and Inequalities

LESSON SUMMARY

This lesson explains linear equations and systems of linear equations and inequalities and shows you how to solve them graphically.

What Is a Linear Equation?

When you graphed linear equations in Lesson 8, you found that their graph was always a straight line. If the graph of an equation is a straight line, the equation is a *linear equation*. Did you notice that the word *line* is part of the word *linear*? That may help you remember that a linear equation will always graph into a straight line.

There are other ways to determine if an equation is linear without graphing the equation. Equations of this type can be put into the form Ax + By = C, where A and B are not both equal to zero. This is called the *standard form* of a linear equation. In standard form, a linear equation can have no exponent greater than 1, cannot have any variables in the denominator if the equation contains both variables, and cannot have the product of variables in the equation.

Here are examples of equations that *are* linear:

$2x + y = 11$
$\frac{2}{3}x + 3y = 12$
$5x - 2y = 16$
$x = 18$

Here are examples of equations that *are not* linear:

$x^2 + y = 5$	Equation contains a variable greater than 1.
$\frac{2}{x} + y = 7$	Equation contains the variable in the denominator.
$xy = 6$	Equation contains the product of two variables.

Practice

Determine if the following equations are linear equations.

______ **1.** $5x + y = 13$

______ **2.** $6x - 3y = 8$

______ **3.** $y = 12$

______ **4.** $x + y^2 = 11$

______ **5.** $3x + 2xy - 4y = 5$

______ **6.** $\frac{3}{4}x + 2y = 5$

______ **7.** $\frac{3}{x} + y = 17$

______ **8.** $y^2 + 5x + 6 = 0$

______ **9.** $x + y^3 = 11$

______ **10.** $3x + 2y + 5 = x - y + 7$

► What Is a System of Linear Equations?

A *system* is two or more equations with the same variables. If you have two different variables, you need two equations. If you have three different variables, you need three equations. There are several methods of solving systems of equations. In this lesson, you will solve systems of equations graphically. When you graphed linear equations in Lesson 8, the graph, which was a straight line, was a picture of the answers; and you had an infinite number of solutions.

However, a system of linear equations has more than one equation, so its graph would be more than one line. You'll know that you've solved a system of linear equations when you determine the point(s) of intersection of the lines. Since two lines can intersect in only one point, that means the system of linear equations has one solution. What if the lines don't intersect? When the lines do not intersect, you have no solution.

Generally speaking, two lines can intersect in only one point, or they cannot intersect at all. However, there is a third possibility. Do you know what it is? The lines could *coincide,* which means they are the same line. If the lines coincide, there are an infinite number of solutions since every point on the line is a point of intersection.

The graphs of a linear system would be one of three cases, as shown on pages 83 and 84.

Case 1

The lines intersect in one point. When the lines intersect in one point, you have one solution.

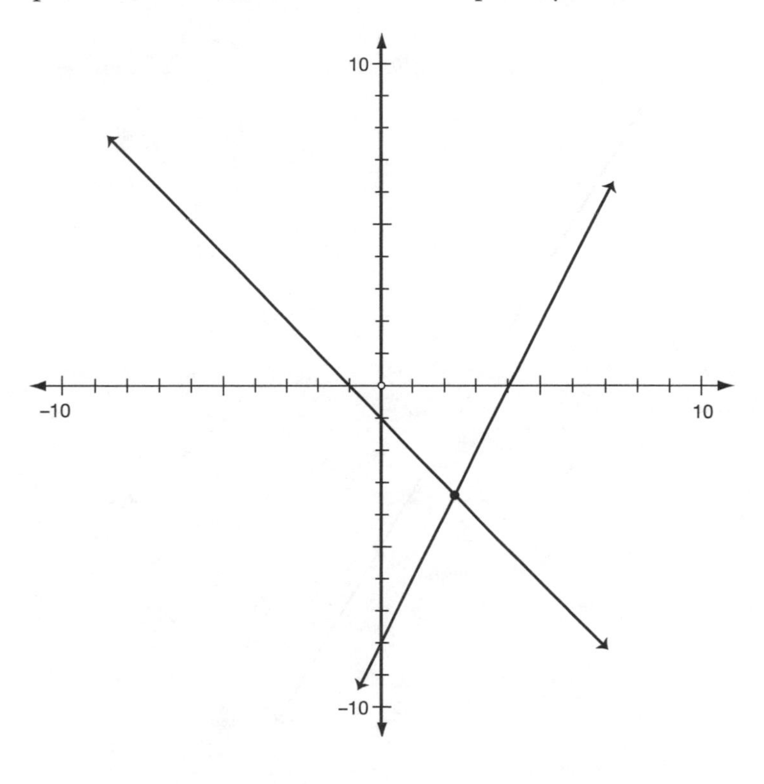

Case 2

The lines do not intersect. When the lines do not intersect, you have no solutions (∅).

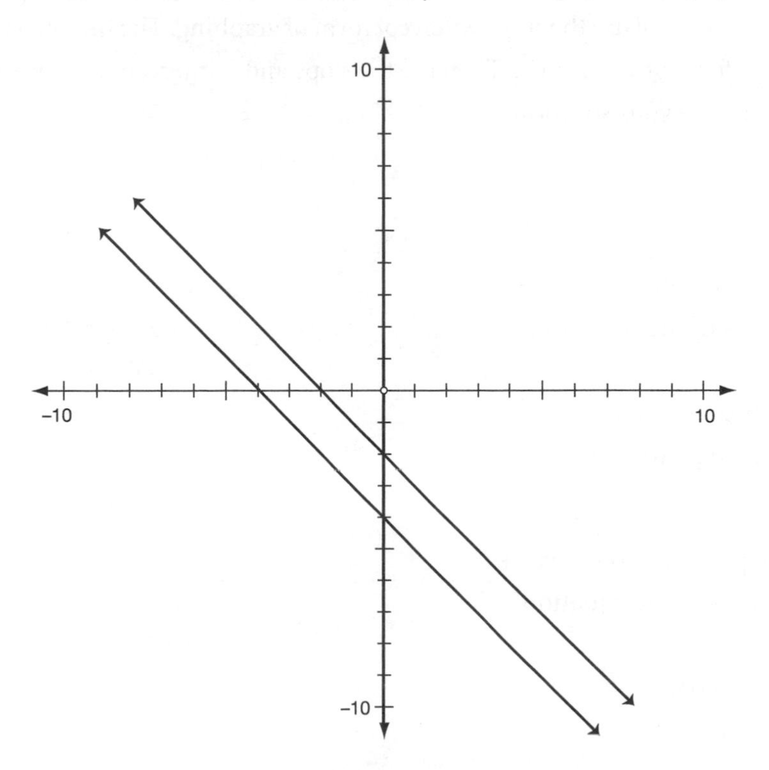

Case 3

The lines coincide. When the lines coincide, you have an infinite number of solutions.

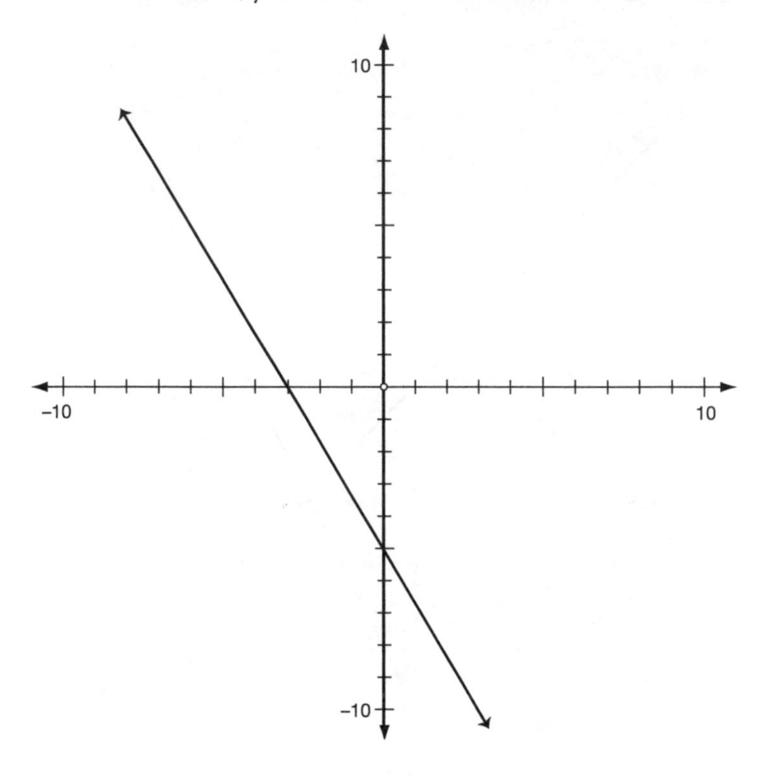

Graphing a System of Linear Equations

To graph a system of linear equations, use what you already know about graphing linear equations. To graph a system of linear equations, you will use the slope-intercept form of graphing. The first step is to transform the equations into slope-intercept form or $y = mx + b$. Then use the slope and y-intercept to graph the line. Once you have both lines graphed, determine your solutions.

Example: $x - y = 6$

$2x + y = 3$

Transform the first equation into $y = mx + b$.	$x - y = 6$
Subtract x from both sides of the equation.	$x - x - y = 6 - x$
Simplify.	$-y = 6 - x$
Use the commutative property.	$-y = -x + 6$
Multiply both sides of the equation by -1.	$-1 \cdot -y = -1(-x + 6)$
Simplify both sides.	$y = x - 6$

Transform the second equation into $y = mx + b$.	$2x + y = 3$
Subtract $2x$ from both sides of the equation.	$2x - 2x + y = 3 - 2x$
Simplify.	$y = 3 - 2x$
Use the commutative property.	$y = -2x + 3$

The slope of the first equation is 1, and the y-intercept is –6. The slope of the second equation is –2, and the y-intercept is 3. In the first equation, the line cuts the y-axis at –6. From that point, go up 1 and right 1. Draw a line through your beginning point and the endpoint—this line can extend as long as you want in both directions since it is endless. In the second equation, the line cuts the y-axis at 3. From that point, go down 2 and to the right 1. Draw a line through your beginning point and the endpoint, extending it as long as you want. The point of intersection of the two lines is (3,–3), so there is one solution.

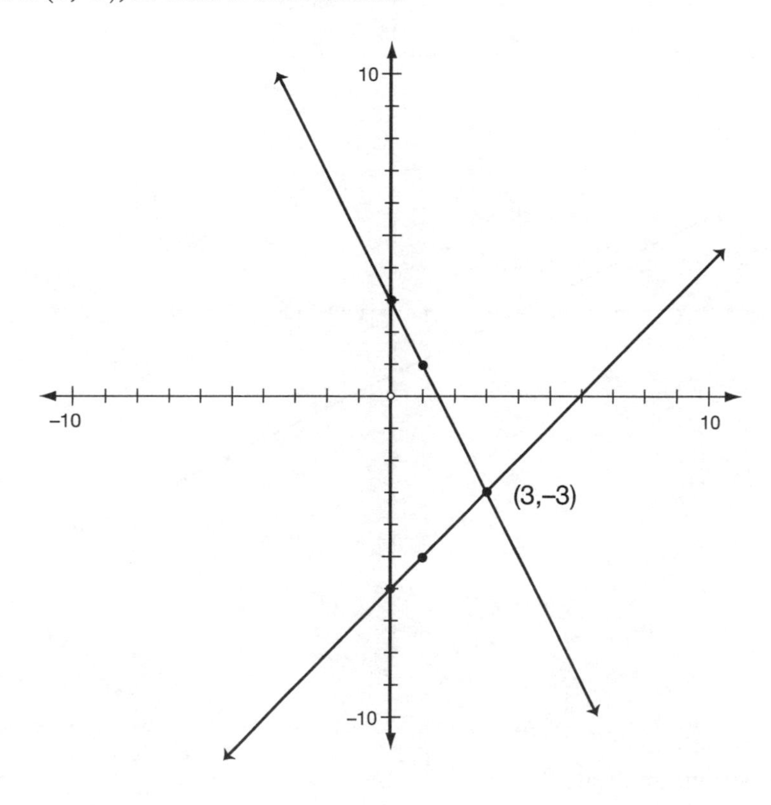

Example: $x + 2y = -4$

$2x + 4y = -8$

Transform the first equation into $y = mx + b$.	$x + 2y = -4$
Subtract x from both sides of the equation.	$x - x + 2y = -4 - x$
Simplify.	$2y = -4 - x$
Use the commutative property.	$2y = -x - 4$
Divide both sides of the equation by 2.	$\frac{2y}{2} = \frac{-x}{2} + \frac{-4}{2}$
Simplify both sides of the equation.	$y = -\frac{1}{2}x - 2$
Transform the second equation into $y = mx + b$.	$2x + 4y = -8$
Subtract $2x$ from both sides of the equation.	$2x - 2x + 4y = -8 - 2x$
Simplify.	$4y = -8 - 2x$
Use the commutative property.	$4y = -2x - 8$
Divide both sides of the equation by 4.	$\frac{4y}{4} = \frac{-2x}{4} - \frac{8}{4}$
Simplify both sides of the equation.	$y = -\frac{1}{2}x - 2$

The first and second equations transformed into the same equation, $y = -\frac{1}{2}x - 2$. If you have the same equation, it will graph into the same line. This is a case where the two lines coincide. Graph the line. Start with the y-intercept, which is –2. From that point, go down 1 and to the right 2. Since the lines coincide, there are an infinite number of solutions.

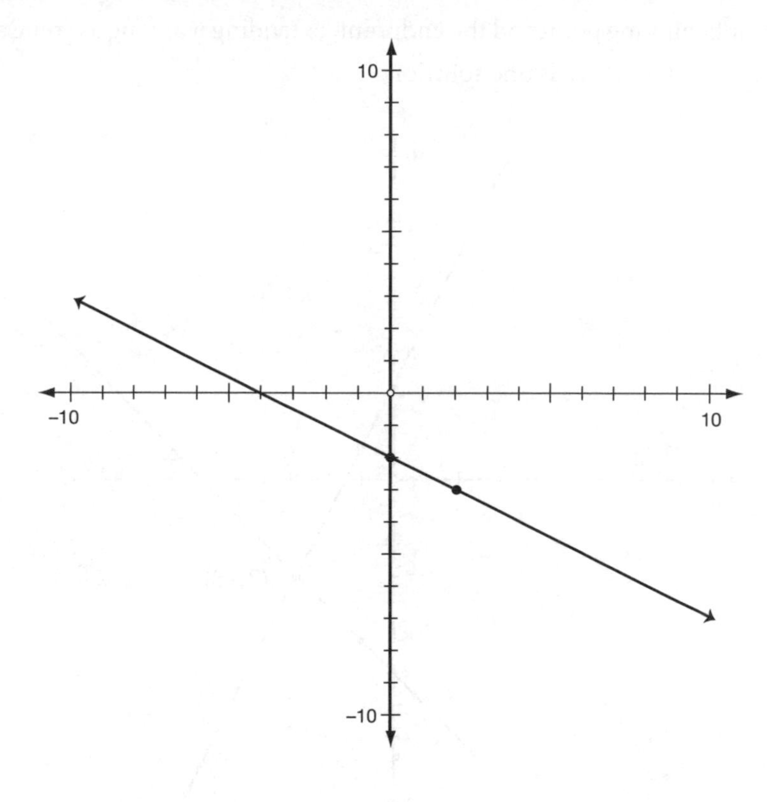

Practice

Graph the systems of equations and state the solution(s).

11. $x + 2y = 8$
$x - y = -4$

12. $x - y = 5$
$3x - 3y = 15$

13. $2x - y = 6$
$2x - y = 1$

14. $y - x + 3 = 3$
$y + 2x = 3$

15. $y = 3x - 1$
$y = 3x + 2$

16. $2y + x = 1$
$y - x = 5$

Here's a time-saver! You can determine the nature of the solutions of a system without actually graphing them. Once you have the equations transformed into $y = mx + b$, compare the slopes and y-intercepts of each equation. If the slopes and y-intercepts are the same, you will have the same line. If the slopes are the same, but the y-intercepts are different, then the lines will be parallel. If the slopes are different, then the lines will intersect.

Case 1: The slopes and y-intercepts are the same, so the lines coincide. There will be an infinite number of solutions.

$y = 2x + 3$
$y = 2x + 3$

Case 2: The slopes are the same, but the y-intercepts are different. The lines will be parallel, so there is no solution.

$y = 2x + 4$
$y = 2x - 5$

Case 3: The slopes are different. The lines will intersect, so there will be one solution.

$y = 3x + 2$
$y = 2x + 3$

Practice

Without graphing the system, determine the number of solutions the system will have.

______**17.** $y = 3x - 5$
$y = 3x + 2$

______**18.** $x + 3y = 10$
$2x + 6y = 20$

______**19.** $3y - 2x = 6$
$2y - 3x = 4$

______**20.** $2x + 3y = 6$
$3x - y = 2$

______**21.** $y + 3 = 3x + 5$
$3y = 9x$

______**22.** $3x + y = 6$
$3x - y = 6$

______**23.** $4x - 3y = 12$
$-4x + 3y = -12$

______**24.** $3x + 3y = 15$
$-2x - 2y = 8$

▶ Solving Systems of Inequalities Graphically

A system of inequalities is two or more inequalities with the same variables. You graph systems of linear inequalities like you graph systems of linear equations. However, remember that the graph of an inequality consists of a boundary line *and* a shaded area. Review Lesson 10 if you need help recalling the basics of graphing inequalities.

To graph a system of inequalities, transform the inequality into $y = mx + b$, and graph the boundary line. Then determine if you will shade above or below the boundary line. The solution of the system of inequalities will be the intersection of the shaded areas.

Example: $y > x$
$y < 3$

The inequalities are already in $y =$ *form,* so you are ready to graph them. The slope of the first equation is 1, and the y-intercept is 0. Start with the y-intercept of 0 and then go up 1 and to the right 1. The boundary line will be dotted because the inequality symbol is >. Because the inequality is >, you will shade above the line.

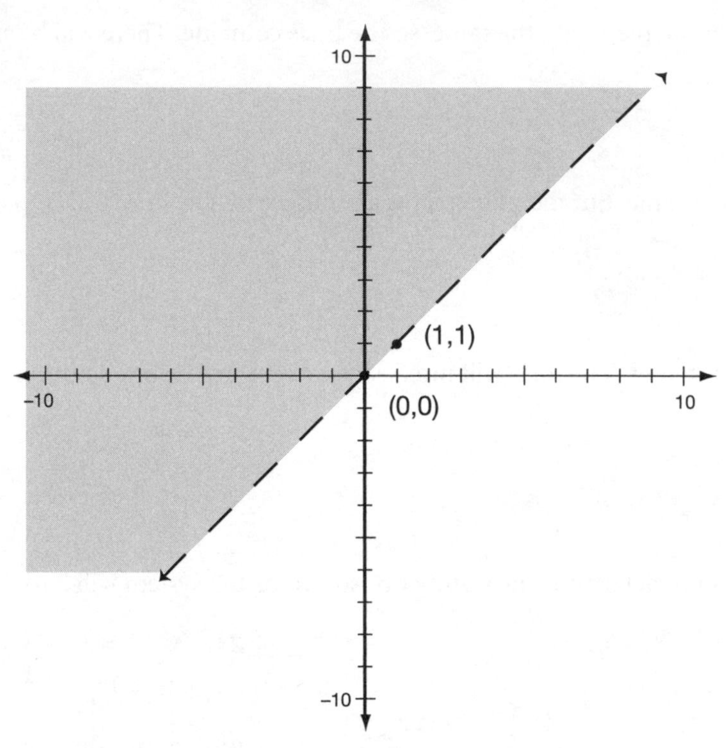

The inequality $y < 3$ will have a horizontal boundary line. The boundary line will be dotted, and you will shade below the line because the inequality symbol is $<$.

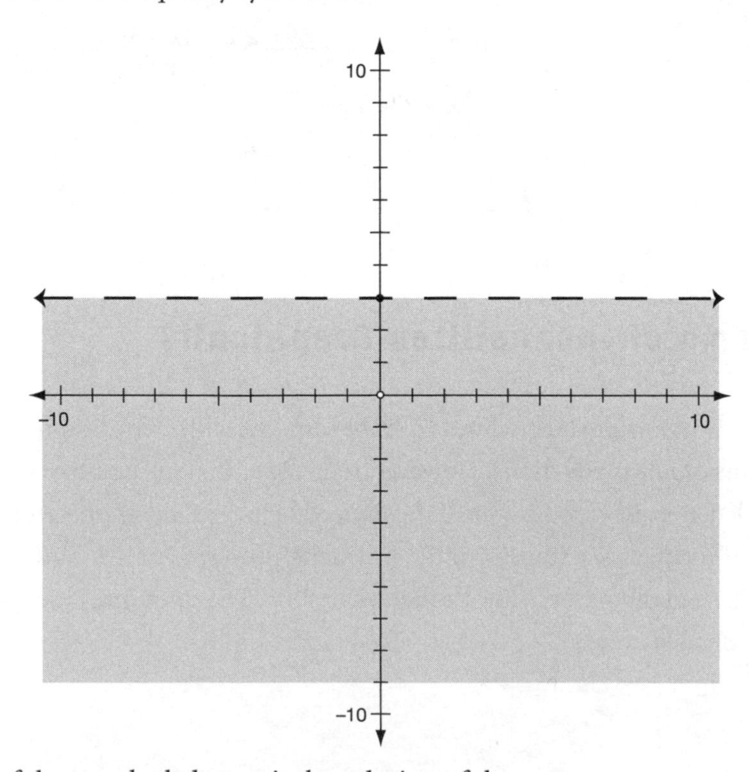

The intersection of the two shaded areas is the solution of the system.

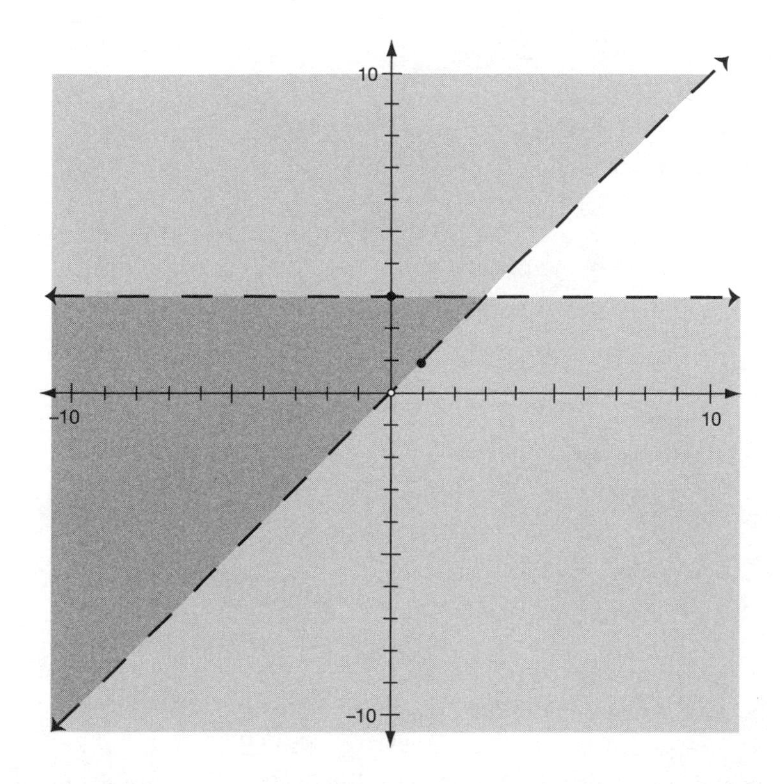

Example: $-x + y \geq 4$

$2x + y \leq 1$

Transform the first inequality to $y = mx + b$.	$-x + y \geq 1$
Add x to both sides of the inequality.	$-x + x + y \geq 4 + x$
Simplify.	$y \geq 4 + x$
Use the commutative property.	$y \geq x + 4$

Transform the second inequality to $y = mx + b$.	$2x + y \leq 1$
Subtract $2x$ from both sides of the inequality.	$2x - 2x + y \leq 1 - 2x$
Simplify.	$y \leq 1 - 2x$
Use the commutative property.	$y \leq -2x + 1$

The slope of the first inequality is 1 and the y-intercept is 4. To graph the inequality, start with the y-intercept, which is 4. From that point, go up 1 and to the right 1. Draw a line through the starting point and the endpoint. The boundary line will be a solid line, and you will shade above it because the inequality symbol is $\geq$.

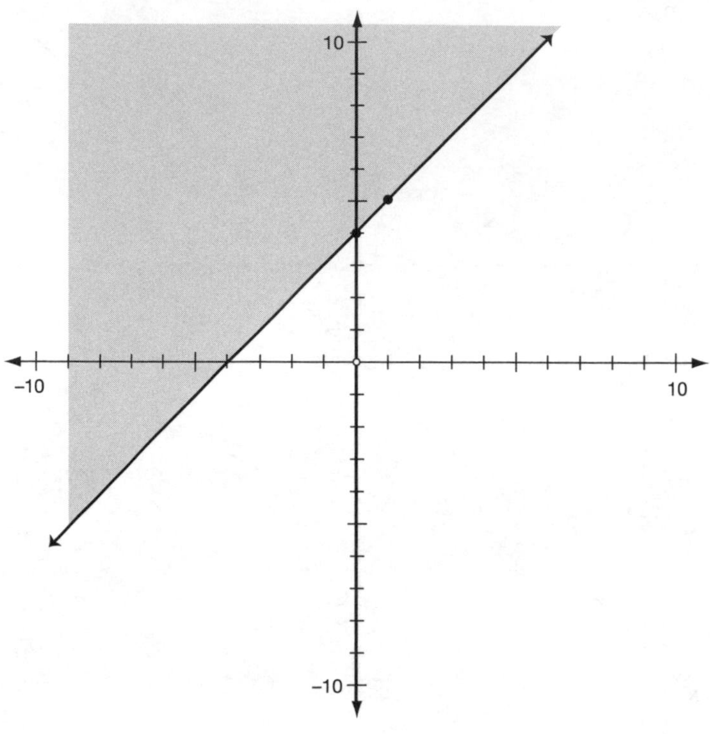

The slope of the second inequality is –2, and the y-intercept is 1. To graph the inequality, start with the y-intercept, which is 1. From that point, go down 2 and to the right 1. Draw a line through the starting point and the endpoint. The boundary line will be a solid line, and you will shade below the line because the inequality symbol is $\leq$.

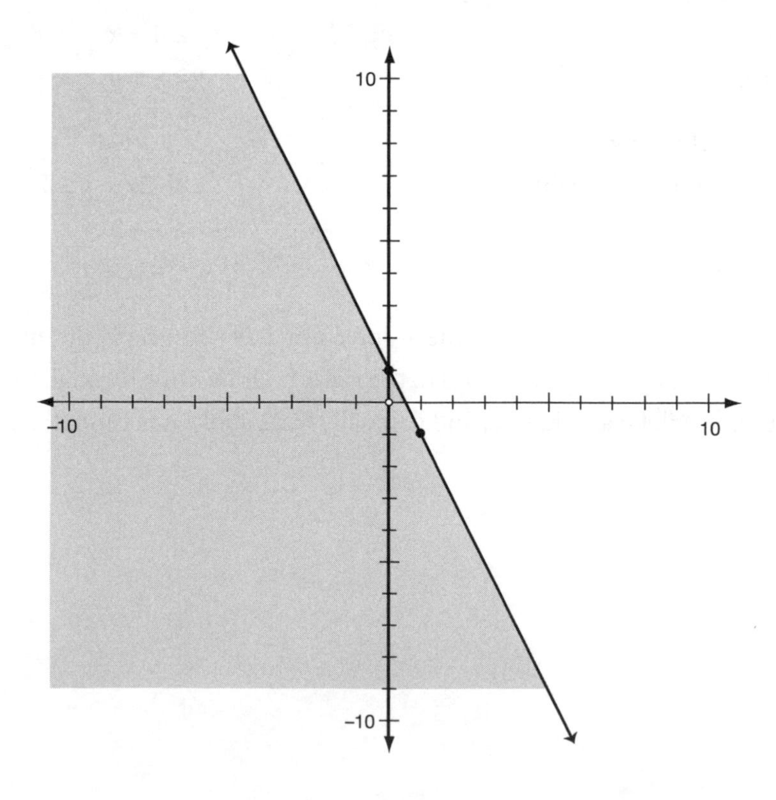

The solution of the system of inequalities is the intersection of the two shaded areas.

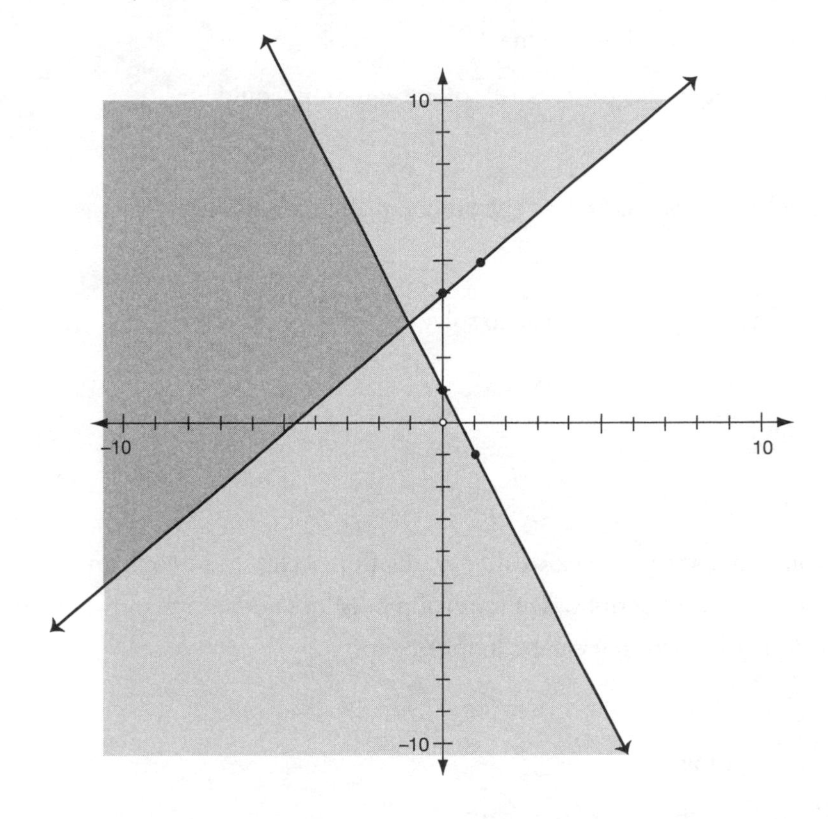

Practice

Solve the systems of inequalities graphically.

25. $y > 4$
$y < x + 2$

26. $y \geq 5$
$x \leq 2$

27. $y < x + 2$
$y < -x + 4$

28. $x + y > 5$
$-2x + y > 3$

29. $x \geq 0$
$y \geq 0$
$y \leq -2x + 5$

30. $y \leq \frac{3}{2}x + 5$
$y < \frac{3}{2} + 2$

Applications

Work through the following word problems carefully.

31. The ticket price of the movie in Cinema 1 is \$5. The ticket price of the movie in Cinema 2 is \$7. Your group of 10 people spent \$62 on tickets. How many people attended Cinema 1, and how many people attended Cinema 2?

Let x = number of persons attending Cinema 1

Let y = number of persons attending Cinema 2

The total number of people in the group is 10, so one equation would be:

$x + y = 10$

The total spent on ticket prices was $62, so the other equation would be:

$5x + 7y = 62$

The system of equations that you would use to solve this problem is:

$x + y = 10$

$5x + 7y = 62$

Solve the system graphically.

32. The senior class is planning a trip to Mexico after graduation. At least 12 people must sign up for the trip, or it will be canceled. The number of boys in the group must not be more than twice the number of girls. What is the least number of girls that can be in the group?

Let x = number of girls going

Let y = number of boys going

There must be at least 12 in the group, so $x + y \geq 12$.

The number of boys must not outnumber the girls by more than twice as much, so $2x \leq y$.

The system of equations that would solve the problem is:

$x + y \geq 12$

$2x \leq y$

Graph the system of inequalities to solve the problem.

Skill Building until Next Time

Use a system of inequalities to represent how you will spend your money. Let x = the amount of money you need to spend on necessities. Let y = the amount of money you can spend on recreation. Fill in the system of inequalities with the amount of money that fits your circumstances.

$x + y \leq$ (amount of money you have to spend)

$x \geq$ (amount of money you need to spend on necessities)

When you have filled in the dollar amounts, graph the system of inequalities. You do not want any values less than zero. Why?

LESSON

12 Solving Systems of Equations Algebraically

LESSON SUMMARY

You know how to solve systems of equations graphically. In this lesson, you will learn how to solve systems of equations algebraically. You will learn how to use the elimination method and the substitution method for solving systems of equations.

▶ How to Use the Elimination Method

Graphs serve many useful purposes, but using algebra to solve a system of equations can be faster and more accurate than graphing a system. A system of equations contains equations with more than one variable. If you have more than one variable, you need more than one equation to solve for the variables. When you use the elimination method of solving equations, the strategy is to eliminate all the variables except one. When you have only one variable left in the equation, then you can solve it.

Example: $x + y = 10$

$x - y = 4$

Add the equations. $2x - 0y = 14$

Drop the $0y$. $2x = 14$

Divide both sides of the equation by 2. $\frac{2x}{2} = \frac{14}{2}$

Simplify both sides of the equation. $x = 7$

You have solved for the variable x. To solve for the variable y, substitute the value of the x variable into one of the original equations. It does not matter which equation you use. $x + y = 10$

Substitute 7 in place of the x variable. $7 + y = 10$

Subtract 7 from both sides of the equation. $7 - 7 + y = 10 - 7$

Simplify both sides of the equation. $y = 3$

You solve a system of equations by finding the value of all the variables. In the previous example, you found that $x = 7$ and $y = 3$. Write your answer as the ordered pair (7,3). To check your solution, substitute the values for x and y into *both* equations.

Check: $x + y = 10$

Substitute the values of the variables into the first equation. $7 + 3 = 10$

Simplify. $10 = 10$

$x - y = 4$

Substitute the values of the variables into the second equation. $7 - 3 = 4$

Simplify. $4 = 4$

Did you get the right answer? Because you got true statements when you substituted the value of the variables into both equations, you solved the system of equations correctly. Try another example.

Example: $x + y = 6$

$-x + y = -4$

Add the two equations. $0x + 2y = 2$

Drop the $0x$. $2y = 2$

Divide both sides of the equation by 2. $\frac{2y}{2} = \frac{2}{2}$

Simplify both sides of the equation. $y = 1$

Use one of the original equations to solve for x. $x + y = 6$

Substitute 1 in place of y. $x + 1 = 6$

Subtract 1 from both sides of the equation. $x + 1 - 1 = 6 - 1$

Simplify both sides of the equation. $x = 5$

Write the solution of the system as an ordered pair: (5,1).

Check: $x + y = 6$

Substitute the values of x and y into the first equation. $5 + 1 = 6$

Simplify. $6 = 6$

$-x + y = -4$

Substitute the values of x and y into the second equation. $-(5) + 1 = -4$

Simplify. $-4 = -4$

Did you get the right answer? Yes! You got true statements when you substituted the value of the variables into both equations, so know you solved the system of equations correctly.

Practice

Solve the systems of equations using the elimination method.

______ **1.** $x + y = 8$
$x - y = 18$

______ **2.** $x - 2y = 4$
$x + 2y = 16$

______ **3.** $-3x + 2y = -12$
$3x + 5y = -9$

______ **4.** $4x + 2y = 12$
$-4x + y = 3$

______ **5.** $2x - 3y = 9$
$x + 3y = 0$

______ **6.** $2x - y = 2$
$-2x + 4y = -5$

What If You Can't Easily Eliminate a Variable?

Sometimes, you can't easily eliminate a variable. Take a look at the following example. What should you do?

Example: $x + y = 24$
$2x + y = 3$

If you were to add this system of equations the way it is, you would be unable to eliminate a variable. However, if one of the *y* variables were negative, you would be able to eliminate the *y* variable. You can change the equation to eliminate the *y* variable by multiplying one of the equations by –1. You can use either equation. Save time by choosing the equation that looks easier to manipulate.

Multiply both sides of the first equation by –1.	$-1(x + y) = -1 \cdot 24$
Simplify both sides of the equation.	$-x - y = -24$
Add the second equation to the modified first equation.	$2x + y = 3$
	$x = -21$
	$x + y = 24$
Substitute the value of *x* into one of the original equations.	$-21 + y = 24$
Add 21 to both sides of the equation.	$-21 + 21 + y = 24 + 21$
Simplify both sides of the equation.	$y = 45$

The solution to the system of equations is (–21,45).

Example: $2x + y = 4$
$3x + 2y = 6$

How will you eliminate a variable in this system? If you multiply the first equation by –2, you can eliminate the *y* variable.

	$-2(2x + y) = -2 \cdot 4$
	$-4x - 2y = -8$
Add the second equation to the modified first equation.	$3x + 2y = 6$
	$-x = -2$
Multiply both sides of the equation by –1.	$-1 \cdot -x = -1 \cdot -2$
	$x = 2$

Substitute 2 in place of the x variable in one of the original equations.

$2x + y = 4$

$2 \cdot 2 + y = 4$

Simplify. $4 + y = 4$

Subtract 4 from both sides of the equation. $4 - 4 + y = 4 - 4$

Simplify. $y = 0$

The solution of the system is (2,0).

Example: $5x - 2y = 8$

$3x - 5y = 1$

Are you ready for a challenge? To eliminate one of the variables in this system of equations, you will need to alter both equations. If you multiply the first equation by 5 and the second equation by –2, then you will be able to eliminate the y variable. Or, you could multiply the first equation by 3 and the second equation by –5 to eliminate the x variable. To decide which variable to eliminate, choose the one that looks like it will be the easier one to do.

Multiply the first equation by 5. $5(5x - 2y) = 8 \cdot 5$

Multiply the second equation by –2. $-2(3x - 5y) = 1 \cdot -2$

Simplify the first equation. $25x - 10y = 40$

Simplify the second equation. $\underline{-6x + 10y = -2}$

Add both equations. $19x = 38$

Divide both sides of the equation by 19. $\frac{19x}{19} = \frac{38}{19}$

Simplify both sides of the equation. $x = 2$

Substitute the 2 in place of the x variable into one of the *original* equations. $5x - 2y = 8$

$5 \cdot 2 - 2y = 8$

Simplify. $10 - 2y = 8$

Subtract 10 from both sides of the equation. $10 - 10 - 2y = 8 - 10$

Simplify both sides of the equation. $-2y = -2$

Divide both sides of the equation by –2. $\frac{-2y}{-2} = \frac{-2}{-2}$

Simplify both sides of the equation. $y = 1$

The solution of the system is (2,1).

Rearranging Equations

Sometimes, it is necessary to rearrange an equation before you can solve it using the elimination method. Take a look at this example.

Example: $2x = 2y + 6$
$3x = -3y + 3$

Rearrange the first equation.	$2x = 2y + 6$
Subtract $2y$ from both sides of the equation.	$2x - 2y = 2y - 2y + 6$
Simplify.	$2x - 2y = 6$
Rearrange the second equation.	$3x = -3y + 3$
Add $3y$ to both sides of the equation.	$3x + 3y = -3y + 3y + 3$
Simplify.	$3x + 3y = 3$
Your system has been altered to these equations.	$2x - 2y = 6$
	$3x + 3y = 3$
Multiply the first equation by 3.	$3(2x - 2y) = 3 \cdot 6$
Multiply the second equation by 2.	$2(3x + 3y) = 2 \cdot 3$
Simplify both equations.	$6x - 6y = 18$
	$\underline{6x + 6y = 6}$
Add both equations.	$12x = 24$
Divide both sides of the equation by 12.	$\frac{12x}{12} = \frac{24}{12}$
Simplify both sides.	$x = 2$
Substitute 2 in place of x in one of the original equations.	$2x = 2y + 6$
	$2 \cdot 2 = 2y + 6$
Simplify.	$4 = 2y + 6$
Subtract 6 from both sides of the equation.	$4 - 6 = 2y + 6 - 6$
Simplify both sides of the equation.	$-2 = 2y$
Divide both sides by 2.	$\frac{-2}{2} = \frac{2y}{2}$
Simplify.	$-1 = y$

The solution of the system is $(2, -1)$.

Practice

Solve the systems of equations using the elimination method.

______ **7.** $x + y = 13$
$x + 2y = 1$

______ **8.** $2x - y = 7$
$4x - y = 9$

______ **9.** $3x + 2y = 10$
$4x + 2y = 2$

______ **10.** $2x + 3y = 23$
$7x - 3y = 13$

______ **11.** $8x - 4y = 16$
$4x + 5y = 22$

______ **12.** $5x - 3y = 31$
$2x + 5y = 0$

______ **13.** $3x - 4y = 6$
$-5x + 6y = -8$

______ **14.** $6x - 2y = -6$
$-5x + 5y = 5$

______ **15.** $-5x + 2y = 10$
$3x + 6y = 66$

______ **16.** $2x + 2 = 3y - x - 4$
$5x - 2y + 1 = 5$

______ **17.** $x = 3y + 6$
$3x = 6y - 15$

______ **18.** $3x + 2y = x - 6$
$3(x + 2y) = 3$

► How to Use the Substitution Method

Some equations are easier to solve using the substitution method instead of the elimination method. Study the following examples.

Example: $y = 2x$
$2x + y = 12$

The first equation says that $y = 2x$. If you substitute $2x$ in place of y in the second equation, you can eliminate a variable. $2x + 2x = 12$

Combine similar terms. $4x = 12$

Divide both sides of the equation by 4. $\frac{4x}{4} = \frac{12}{4}$

Simplify both sides of the equation. $x = 3$

Substitute 3 in place of x in one of the original equations. $y = 2x$

$y = 2 \cdot 3$

$y = 6$

The solution of the system is (3,6).

Example: $x = 2y + 1$
$x + 3y = 16$

Substitute $2y + 1$ in place of x in the second equation.	$2y + 1 + 3y = 16$
Simplify.	$5y + 1 = 16$
Subtract 1 from both sides of the equation.	$5y + 1 - 1 = 16 - 1$
Simplify both sides of the equation.	$5y = 15$
Divide both sides of the equation by 5.	$\frac{5y}{5} = \frac{15}{5}$
Simplify both sides of the equation.	$y = 3$
Substitute 3 in place of y in one of the original equations.	$x = 2y + 1$
	$x = 2 \cdot 3 + 1$
	$x = 7$

The solution for the system is (7,3).

Practice

Solve the systems of equations using the substitution method.

______**19.** $x = 3y$
$2x + y = 14$

______**20.** $x = -5y$
$2x + 2y = 16$

______**21.** $y = 4x$
$3x + 4y = 38$

______**22.** $y = 2x + 1$
$3x + 2y = 9$

______**23.** $x = 2y + 1$
$3x - y = 13$

______**24.** $y = 3x + 2$
$2x - 3y = 8$

Mixed Practice

Solve the systems of equations using any method.

______**25.** $x + 2y = 10$
$2x - 2y = -4$

______**26.** $y = -6x$
$2x - y = 16$

______**27.** $4x + 2y = 18$
$3x - y = 1$

______**28.** $x = 5y - 2$
$2x - y = 14$

______**29.** $3x + 7y = 4$
$6x + 2y = -4$

______**30.** $2x - 5y = 10$
$3x + 2y = -4$

Applications

Use a system of equations to solve the following word problems. First, work through the example problem given.

Example: Your class is having a pizza party. The pepperoni pizza is \$8, and the combination pizza is \$12. You need 9 pizzas, and you have \$80 to spend. How many of each kind can you get?

Let x = number of pepperoni pizzas

Let y = number of combination pizzas

	$x + y = 9$
	$8x + 12y = 80$
Multiply the first equation by –8.	$-8(x + y) = -8 \cdot 9$
Simplify.	$-8x - 8y = -72$
Add the second equation to the altered first equation.	$\underline{8x + 12y = 80}$
	$4y = 8$
Divide both sides of the equation by 4.	$\frac{4y}{4} = \frac{8}{4}$
Simplify both sides of the equation.	$y = 2$
Substitute 2 in place of y in one of the original equations.	$x + y = 9$
	$x + 2 = 9$
Subtract 2 from both sides of the equation.	$x + 2 - 2 = 9 - 2$
Simplify.	$x = 7$

You can get 7 pepperoni pizzas and 2 combination pizzas.

31. The admission prices for the last baseball game of the season were \$3 for students and \$4 for adults. One hundred tickets were sold, and the gate receipts were \$340. How many of each kind of tickets were sold?

32. Jolynn drove 370 miles in 6 hours. She drove 70 miles per hour on the highway and 45 miles per hour on the back roads. How many hours did she spend on the back roads?

Skill Building until Next Time

What kind of problems in your everyday life could be solved using a system of equations? Use the problems listed under the *Applications* heading to help you generate ideas. After you come up with an idea for a system of equations, go ahead and solve it using the elimination or substitution method. Then, check your answer as shown in this lesson to be sure you got the correct answer.

LESSON

13 Working with Exponents

LESSON SUMMARY

This lesson shows you how to add, subtract, multiply, and divide expressions with exponents. You will also learn how to raise expressions to a power.

What Is an Exponent?

An *exponent* tells you how many times a factor is multiplied. An exponent appears as a raised number that is smaller in size than the other numbers on the page. For example, in the expression 4^3, the three is the exponent. The expression 4^3 shows that four is a factor three times. That means four times four times four. Here are examples of exponents and what they mean:

$5^2 = 5 \cdot 5$

$2^3 = 2 \cdot 2 \cdot 2$

$a^2 = a \cdot a$

$2x^3y^2 = 2 \cdot x \cdot x \cdot x \cdot y \cdot y$

▶ Adding and Subtracting with Exponents

In previous lessons, when you combined similar terms, you added the numbers in front of the variables (coefficients) and left the variables the same. Here are some examples:

$3x + 4x = 7x$
$2x^2 + 7x^2 = 9x^2$
$3xy + 6xy = 9xy$
$5x^3 - 3x^3 = 2x^3$

What do you do with exponents when you are adding? Nothing! That's right, you add only the coefficients. The variables and their exponents stay the same. This is not a new concept. You used it in previous lessons when you combined similar terms.

▶ Multiplying with Exponents

The rules for multiplying expressions with exponents may appear to be confusing. You treat exponents differently from ordinary numbers. You would think that when you are multiplying, you would multiply the exponents. However, that's not true. When you are multiplying expressions, you *add* the exponents. Here's an example of how to simplify an expression.

Example: $x^2 \cdot x^3$
$(x \cdot x)(x \cdot x \cdot x)$
x^5

You can see that you have 5 x's, which is written as x^5. To get x^5 for an answer, you *add* the exponents instead of multiplying them.

Tip

When an expression is written as x^2x^5, it implies multiplication. You do not need to use the multiplication symbol.

Example: $a^3 \cdot a^4$
$(a \cdot a \cdot a)(a \cdot a \cdot a \cdot a)$
a^7

The factors of a^3 are $a \cdot a \cdot a$. The factors of a^4 are $a \cdot a \cdot a \cdot a$. The factored form of $a^3 \cdot a^4$ is $a \cdot a \cdot a \cdot a \cdot a \cdot a \cdot a$. When you write the problem out in factor form, you can see that you have 7 a's. The easy way to get 7 is to add the exponents.

What would you do if you see an expression like x^{20}, and you want to multiply it by x^{25}? You can see that writing out the factors of $x^{20} \cdot x^{25}$ would take a long time. Think about how easy it would be to make a mistake if you wrote out all the factors. It is much more efficient and fast to use the rule for multiplying exponents: When you are multiplying, you add the exponents.

What Is a Base?

In the expression x^4, the x is the *base,* and the 4 is the exponent. You can multiply variables only when the base is the same. You can multiply $a^2 \cdot a^4$ to get a^6. However, you cannot multiply $a^3 \cdot b^4$ because the bases are different.

To multiply $2x^2 \cdot 3x^4$, multiply the numbers, keep the variable the same, and add the exponents. Your answer would be $6x^6$. When you multiply $5a^3 \cdot 2a^4$, you get $10a^7$.

Tip

If there is no exponent, the exponent is understood to be 1. For example: $x \cdot x^3 = x^1 \cdot x^3 = x^4$.

Practice

Simplify the expressions using the rules for adding and multiplying with exponents.

_____ **1.** $11x + 17x$

_____ **2.** $3x + 4x^2$

_____ **3.** $a^7 \cdot a^3$

_____ **4.** $x^2y + 5x^2y$

_____ **5.** $3x \cdot 4x^2$

_____ **6.** $ab^2 \cdot a^2b^3$

_____ **7.** $2m^5n^2 \cdot 6mn^9$

_____ **8.** $2r^2s^5 \cdot 3rs^2$

_____ **9.** $3ab^3c \cdot 5a^3bc$

_____ **10.** $3x^2y + 4x \cdot 5xy$

▶ Dividing with Exponents

Now you know that when you multiply expressions with exponents, you add the exponents. However, when you divide, you *subtract* the exponents. Take a look at some examples.

Examples: $\frac{x^5}{x^2} = \frac{x \cdot x \cdot x \cdot x \cdot x}{x \cdot x} = x^{5-2} = x^3$

$\frac{x^2}{x^5} = \frac{x \cdot x}{x \cdot x \cdot x \cdot x \cdot x} = \frac{1}{x^{5-2}} = \frac{1}{x^3}$

$\frac{a^3b^2}{ab^5} = \frac{a \cdot a \cdot a \cdot b \cdot b}{a \cdot b \cdot b \cdot b \cdot b \cdot b} = \frac{a^{3-1}}{b^{5-2}} = \frac{a^2}{b^3}$

Tip

When subtracting exponents, always subtract the larger exponent from the smaller exponent. If the larger exponent is in the numerator, the variable and exponent will be in the numerator. For example, $\frac{x^5}{x^2} = x^3$. If the larger exponent is in the denominator, the variable and exponent will be in the denominator. For example, $\frac{x^2}{x^5} = \frac{1}{x^3}$.

Writing out the factors when you work problems takes too long. So, you can use the rule for exponents when you divide. When you divide expressions that contain exponents like $\frac{4x^5}{2x^3}$, divide the numbers (4 and 2) and subtract the exponents (5 and 3). The answer for the expression is $\frac{4x^5}{2x^3} = 2x^2$.

Practice

Simplify the expressions.

______ **11.** $\frac{y^8}{y^3}$

______ **12.** $\frac{a^3}{a^6}$

______ **13.** $-\frac{b^7}{b^3}$

______ **14.** $\frac{a^4b}{a^2}$

______ **15.** $\frac{10xy^2}{2xy}$

______ **16.** $\frac{2ab^3}{18a^3b^7}$

______ **17.** $\frac{16xy^5z^4}{4y^2z^6}$

______ **18.** $\frac{25r^5s^3}{5r^6s}$

______ **19.** $\frac{8x^3y^7}{10xy^{10}}$

______ **20.** $\frac{3m^7n^4}{9m^4n^4}$

Tip

A quantity raised to the zero power is 1. Example: $\frac{x^3}{x^3} = x^0$. However, a number divided by itself is always 1. Therefore, x^0 must equal 1.

▶ What Do You Do with Exponents When You Raise a Quantity to a Power?

How do you simplify the expression $(x^3)^2$? Remember that an exponent tells you how many times a quantity is a factor.

Examples: $(x^3)^2 = (x \cdot x \cdot x)(x \cdot x \cdot x) = x^6$

$(a^2b^3)^2 = (a \cdot a \cdot b \cdot b \cdot b)(a \cdot a \cdot b \cdot b \cdot b) = a^4b^6$

$(3a^3)^2 = (3 \cdot a \cdot a \cdot a)(3 \cdot a \cdot a \cdot a) = 3^2a^6 = 9a^6$

From the previous examples, you can see that if you multiply the exponents, you will get the correct answer. *If you raise a quantity to a power, you multiply the exponents.* Note that if a number has no exponent, the exponent is understood to be 1.

Practice

Simplify the expressions.

______**21.** $(x^5)^2$

______**22.** $(c^4)^5$

______**23.** $(a^2b^3)^2$

______**24.** $(xy^5)^3$

______**25.** $(m^6n^2)^3$

______**26.** $(2x^3)^3$

______**27.** $(3a^4b^2c)^2$

______**28.** $(3xy^5z^{11})^3$

______**29.** $(4c^5d^6)^2$

______**30.** $(2a^3bc^5)^4$

Mixed Practice

Simplify the expressions.

______**31.** $7x^5 + 6x^5$

______**32.** $7x^5 \cdot 6x^5$

______**33.** $2x^3 \cdot 3y^2$

______**34.** $4x \cdot 2xy^3 + 10x^2y^3$

______**35.** $(x^2y^3)^3 \cdot 2x^3$

______**36.** $2x^2 \cdot 3y^5 + 4xy^3 \cdot 2xy^2$

______**37.** $(3a^4b^5)^3$

______**38.** $(2xy^5)^3 - 8xy^8(x^2y^7)$

______**39.** $\frac{12x^3y^2}{2xy} + 5xy^2$

______**40.** $\frac{3a^2b^3 \cdot 2a^4b^7}{8a^5b^{12}}$

Skill Building until Next Time

You may have heard someone refer to an increase as an exponential increase. When something increases exponentially, it means that something increases very rapidly. One example of this type of increase is the growth of cancer cells. You start with one cell. That cell divides to form two cells. Each of the two cells divide, and then you have four cells. This process can be explained using exponents.

$2^0 = 1$

$2^1 = 2$

$2^2 = 4$

$2^3 = 8$

$2^4 = 16$

$2^5 = 32$

$2^6 = 64$

$2^7 = 128$

You can see the rapid increase, which explains why cancer is so devastating.

What would be another real-world example of exponential growth?

LESSON

14 Multiplying Polynomials

LESSON SUMMARY

In this lesson, you will multiply polynomials. The polynomial may have one or more terms. You will also learn a process that is often referred to as "FOIL."

What Is a Polynomial?

A *polynomial* is a number, a variable, or a combination of a number and a variable. A polynomial can be one or more terms. Monomials, binomials, and trinomials are all polynomials. In Lesson 13, you multiplied polynomials with one term when you multiplied $2x^2 \cdot 3xy^3$. In this lesson, you will multiply polynomials that have more than one term.

Multiplying a Polynomial by a Monomial

A polynomial with one term is called a *monomial*. To multiply a polynomial with one term (monomial) by a polynomial with more than one term, use the distributive property. You multiply the term outside the parentheses by every term inside the parentheses.

Examples: $2(a + b - 3) = 2a + 2b - 6$

$3x(x^2 + 2x) = 3x^3 + 6x^2$

Practice

______ **1.** $5(x - y + 2)$

______ **2.** $7x(x - 3)$

______ **3.** $8x^3(3x^2 + 2x - 5)$

______ **4.** $-6(x - y - 7)$

______ **5.** $3b(x^2 + 2xy + y)$

______ **6.** $-7c^2(2a - 5ac)$

______ **7.** $(x^2y - x)6y$

______ **8.** $2a^2x(3x - 2ab + 10ax^3 + 8)$

______ **9.** $-8xy^2(2x^3 - 3x^2y)$

______ **10.** $4rs(-2rt + 7r^2s - 9s^2t^2)$

▶ Multiplying a Binomial by a Binomial

What is a binomial? A *binomial* is a polynomial with two terms. To multiply a binomial by a binomial, you will use a method called "FOIL." This process is called FOIL because you work the problem in this order:

FIRST
OUTER
INNER
LAST

Example: $(x + 2)(x + 3)$

Multiply the **first** terms in each binomial.	$([x] + 2)([x] + 3)$
	$= x^2$
Multiply the two **outer** terms in each binomial.	$([x] + 2)(x + [3])$
	$= x^2 + 3x$
Multiply the two **inner** terms in each binomial.	$(x + [2])([x] + 3)$
	$= x^2 + 3x + 2x$
Multiply the two **last** terms in each binomial.	$(x + [2])(x + [3])$
	$= x^2 + 3x + 2x + 6$
Simplify.	$= x^2 + 5x + 6$

Example: $(x + 3)(x - 1)$

Multiply the two **first** terms.	$([x] + 3)([x] - 1)$
	$= x^2$
Multiply the two **outer** terms.	$([x] + 3)(x - [1])$
	$= x^2 - 1x$
Multiply the two **inner** terms.	$(x + [3])([x] - 1)$
	$= x^2 - 1x + 3x$
Multiply the two **last** terms.	$(x + [3])(x - [1])$
	$= x^2 - 1x + 3x - 3$
Simplify.	$= x^2 + 2x - 3$

Example: $(2x + 1)(3x - 2)$

Multiply the two **first** terms. $([2x] + 1)([3x] - 2)$
$= 6x^2$

Multiply the two **outer** terms. $([2x] + 1)(3x - [2])$
$= 6x^2 - 4x$

Multiply the two **inner** terms. $(2x + [1])([3x] - 2)$
$= 6x^2 - 4x + 3x$

Multiply the two **last** terms. $(2x + [1])(3x - [2])$
$= 6x^2 - 4x + 3x - 2$

Simplify. $= 6x^2 - x - 2$

Practice

_______**11.** $(x + 2)(x + 4)$

_______**12.** $(x + 6)(x - 3)$

_______**13.** $(x - 6)(x - 2)$

_______**14.** $(x - 1)(x + 10)$

_______**15.** $(x + 1)(x + 1)$

_______**16.** $(2x + 1)(x + 5)$

_______**17.** $(3x - 1)(3x + 1)$

_______**18.** $(4x + 2)(5x - 1)$

_______**19.** $(a + b)(a + b)$

_______**20.** $(a - b)(a + b)$

_______**21.** $(3x + 2)(4x - 3y)$

_______**22.** $(2x + y)(x - y)$

_______**23.** $(5x - 2)(6x - 1)$

_______**24.** $(a + b)(c + d)$

_______**25.** $(7y - 2)(7y + 2)$

_______**26.** $(4x - 1)(4x - 1)$

▶ Multiplying a Binomial by a Trinomial

A bicycle has two wheels, a tricycle has three wheels. Likewise, a binomial has two terms, and a trinomial has three terms. Here's how you would multiply a binomial by a trinomial.

Example: $(x + 2)(x^2 + 2x + 1)$

To work this problem, you need to multiply each term in the first polynomial with each term in the second polynomial. You are doing exactly that when you use FOIL.

Multiply x by each term in the second polynomial. $x(x^2 + 2x + 1)$
$= x^3 + 2x^2 + x$

Multiply 2 by each term in the second polynomial. $2(x^2 + 2x + 1)$
$= 2x^2 + 4x + 2$

Simplify. $x^3 + 2x^2 + x + 2x^2 + 4x + 2$
$= x^3 + 4x^2 + 5x + 2$

Example: $(x + 1)(x + 1)(x + 1)$

To work this problem, you need to multiply the first two factors. You will then multiply the result by the third factor.

Multiply	$(x + 1)(x + 1)$
	$= x^2 + x + x + 1$
Simplify.	$= x^2 + 2x + 1$
Then multiply.	$(x^2 + 2x + 1)(x + 1)$
Multiply.	$x^2(x + 1)$
	$= x^3 + x^2$
Multiply.	$2x(x + 1)$
	$= 2x^2 + 2x$
Multiply.	$1(x + 1)$
	$= x + 1$
Simplify.	$= x^3 + x^2 + 2x^2 + 2x + x + 1$
	$= x^3 + 3x^2 + 3x + 1$

Practice

______ **27.** $(x - 2)(x^2 - 2x + 1)$

______ **28.** $(2x + 3)(x^2 + 2x + 5)$

______ **29.** $(x + 3)(x + 2)(x - 2)$

______ **30.** $(3x + 2)(x - 1)(x + 3)$

Skill Building until Next Time

Measure one of the rooms in your home. Write down how you would express the new area of the room if you increased each side of the room by the same amount. Then multiply using the FOIL method.

LESSON

15 Factoring Polynomials

LESSON SUMMARY

This lesson teaches you how to factor algebraic expressions. You will learn how to factor polynomials using the greatest common factor, the difference of two squares, and the trinomial method.

What Is Factoring?

Factoring is the opposite of multiplying. It undoes what multiplication does. When you factor an algebraic expression, you end up with numbers called *factors* that, when multiplied, will give you the original expression. In Lesson 13, you learned the rules for exponents to enable you to multiply polynomials that contain exponents. You will use these same skills to factor expressions. These skills are also used to solve quadratic equations, and equations are the tools that can help you solve real-life problems.

What are the factors of the number 6? Factors are the numbers you multiply to get 6. The number 6 has two sets of factors: $1 \cdot 6$ and $2 \cdot 3$.

Finding the Greatest Common Factor

The first type of factoring you will learn is the *greatest common factor* method. With this method, you look for the greatest factor of two or more numbers.

Example: What is the greatest common factor of 12 and 24?

Factors of 12 are: 1, 2, 3, 4, 6, 12

Factors of 24 are: 1, 2, 3, 4, 6, 8, 12, 24

The greatest factor both numbers have is 12, so the greatest common factor of 12 and 24 is 12.

Example: What is the greatest common factor of 18 and 27?

Factors of 18 are: 1, 2, 3, 6, 9, 18

Factors of 27 are: 1, 3, 9, 27

The greatest factor both numbers have is 9, so the greatest common factor of 18 and 27 is 9.

Another approach to finding the greatest common factor of two expressions is to factor a number until all its factors are prime factors. *Prime factors* are factors that cannot be factored further. The factors of a prime number are 1 and the number itself. The number 5 is prime because its only factors are 1 and 5.

Example: What is the greatest common factor of 12 and 24?

The prime factors of 12 are: $2 \cdot 2 \cdot 3$

The prime factors of 24 are: $2 \cdot 2 \cdot 2 \cdot 3$

The factors that both 12 and 24 have in common are $2 \cdot 2 \cdot 3$, which equals 12. So you know that the greatest common factor of 12 and 24 is 12.

Example: What is the greatest common factor of $8x^3y$ and $24x^2y^2$?

The prime factors of $8x^3y$ are: $2 \cdot 2 \cdot 2 \cdot x \cdot x \cdot x \cdot y$

The prime factors of $24x^2y^2$ are: $2 \cdot 2 \cdot 2 \cdot 3 \cdot x \cdot x \cdot y \cdot y$

The factors that both numbers have in common are $2 \cdot 2 \cdot 2 \cdot x \cdot x \cdot y$, which equals $8x^2y$.

Example: What is the greatest common factor of $30a^2b$ and $40a^3b^3$?

The prime factors of $30a^2b$ are: $2 \cdot 3 \cdot 5 \cdot a \cdot a \cdot b$

The prime factors of $40a^3b^3$ are: $2 \cdot 2 \cdot 2 \cdot 5 \cdot a \cdot a \cdot a \cdot b \cdot b \cdot b$

Both numbers have $2 \cdot 5 \cdot a \cdot a \cdot b$ in common, so the greatest common factor is $10a^2b$.

► Factoring Using the Greatest Common Factor Method

To factor an expression like $2x + 6$, find the greatest common factor of both terms. In this expression, the greatest common factor is 2. To factor the expression, put the greatest common factor outside the parentheses and put what is left inside the parentheses.

Examples: $2x + 6 = 2(x + 3)$

When you use the distributive property on $2(x + 3)$, you will get the original expression, $2x + 6$. You can check your answer by multiplying the factors. You should get the expression you started with, or you did not factor it correctly.

You can factor the expression, $6x + 18$ in two ways:

$6x + 18 = 2(3x + 9)$

$6x + 18 = 6(x + 3)$

What is the correct answer? Remember that you are factoring using the greatest common factor method. The greatest common factor is 6, so the correct answer is $6(x + 3)$.

Practice

Factor using the greatest common factor method.

_____ **1.** $3x + 12$

_____ **2.** $11a + 33b$

_____ **3.** $3a + 6b + 12c$

_____ **4.** $2x^2 + 3x$

_____ **5.** $5x + 9$

_____ **6.** $16x^2 + 20x$

_____ **7.** $x^2y + 3x$

_____ **8.** $8x^3 - 2x^2 + 4x$

_____ **9.** $-6x^3 + 18x^2y$

_____ **10.** $10x^4y^2 - 50x^3y + 70$

_____ **11.** $6a^2 - 39ab$

_____ **12.** $12a^3b^2c + 4a^5bc^2$

_____ **13.** $22x^4y + 55x^2y^2$

_____ **14.** $8x^2 + 12x + 20$

_____ **15.** $5f^3 - 15f + 25$

_____ **16.** $30a^3b^2 + 20a^2b + 35a^4b^2$

▶ Factoring Using the Difference of Two Squares Method

The second type of factoring is the difference of two squares. You will find this method easy to do. But before you learn this method, you need to review the concept of squares. What is a square? A number multiplied by itself equals a *perfect square*. Examples of perfect squares are:

4 because $2 \cdot 2 = 4$

9 because $3 \cdot 3 = 9$

25 because $5 \cdot 5 = 25$

a^2 because $a \cdot a = a^2$

$16b^2$ because $4b \cdot 4b = 16b^2$

d^{10} because $d^5 \cdot d^5 = d^{10}$

Tip

Any even numbered exponent is a perfect square because when you multiply, you add exponents. Example: $c^{12} = c^6 \cdot c^6$.

The method of factoring using the difference of two squares is an easy pattern to remember. The pattern is $(x + y)(x - y)$, where x is the square root of the first term and y is the square root of the second term.

Example: $x^2 - 4$

Both x^2 and 4 are perfect squares, and they are connected by a subtraction sign, which is why the expression is called the difference of two squares. To factor the expression, take the square root of the first term and the square root of the second term. Write it like this: $(x + 2)(x - 2)$

Tip

The sum of two squares cannot be factored. Example: $x^2 + 4$ is prime. It cannot be factored.

Here are two more examples:

Example: $y^2 - 9$
$= (y + 3)(y - 3)$

Example: $16a^2 - 25b^2$
$= (4a + 5b)(4a - 5b)$

Tip

The factors $(x + 6)(x - 6)$ are the same as $(x - 6)(x + 6)$. The *commutative property* says you can change the order of the factors when you multiply and still get the same result. Example: $2 \cdot 3 = 3 \cdot 2$.

Practice

Factor using the difference of two squares method.

______**17.** $a^2 - 49$

______**18.** $b^2 - 121$

______**19.** $4x^2 - 9$

______**20.** $x^2 + y^2$

______**21.** $r^2 - s^2$

______**22.** $366^2 - 100$

______**23.** $a^6 - b^6$

______**24.** $y^2 - 64$

______**25.** $4x^2 - 1$

______**26.** $25x^2 - 4y^2$

______**27.** $x^{25} - 1$

______**28.** $x^4 - 16$

______**29.** $b^{10} - 36$

______**30.** $16a^2 - 25b^2$

▶ Factoring Using the Trinomial Method

You may have found that the process of factoring has been easy so far. However, you will find that using the trinomial method to factor is considerably more complex. This section will challenge you, but you can do it!

What is a trinomial? A *trinomial* is an algebraic expression with three terms. You must have three terms in your expression before you can consider using the trinomial method of factoring.

When the Sign in Front of the Third Term Is Positive

Here are some examples of how you factor using the trinomial method when the sign in front of the third term is positive.

Example: $x^2 + 2x + 1$

The pattern for the factors of a trinomial looks like this:

$$([\quad] \pm [\quad])([\quad] \pm [\quad])$$

To solve this problem, you need to determine what the missing numbers are and whether the problem needs a positive or negative sign. To begin, look at the first term in the trinomial, which is x^2. What are the factors of x^2? You only have one choice: $x \cdot x$, so you will place an x in the first term of each factor.

$$(x \pm [\quad])(x \pm [\quad])$$

Now look at the third term in the trinomial. The third term is 1, and the only factors of 1 are $1 \cdot 1$. So, place 1 in the last term of each factor.

$$(x \pm 1)(x \pm 1)$$

You are almost finished, but you still need to determine the signs of the factors. Look at the original trinomial. The sign of the third term is +1. You know that $+1 \cdot +1 = +1$ and $-1 \cdot -1 = +1$, but only one will give you the correct answer. Look at the original trinomial again. If the sign in front of the second term is positive, you want $+1 \cdot +1$.

$$(x + 1)(x + 1)$$

Now let's use FOIL to check this answer. $(x+1)(x+1) = x^2 + x + x + 1 = x^2 + 2x + 1$

After using FOIL, your result was the trinomial you started with, so you know you factored the trinomial correctly.

Example: $x^2 + 6x + 9$ $([\quad] \pm [\quad])([\quad] \pm [\quad])$

The only choice for factors of x^2 is $x \cdot x$. $(x \pm [\quad])(x \pm [\quad])$

You have two choices for the factors of 9: $1 \cdot 9$ or $3 \cdot 3$. Determining which set of factors you need is a trial-and-error process. However, after you have practiced factoring trinomials, you will find that your intuition will point you in the right direction and may save you some time. First, try using 1 and 9 as the factors.

$(x \pm 1)(x \pm 9)$

Next, determine the sign. Look at the third term in the trinomial, which is +9. To get a +9, the signs would be + times + or – times –. Since the sign in front of the second term is positive, you will use the positive signs.

$(x+1)(x+9)$

Check the answer using FOIL: $(x+1)(x+9) = x^2 + 9x + 1x + 9 = x^2 + 10x + 9$.

The answer does not check out, so you have to try the other factors, 3 and 3.

$(x+3)(x+3) = x^2 + 3x + 3x + 9 = x^2 + 6x + 9$

This is the original trinomial, so you know you have the correct answer.

Example: $x^2 - 7x + 10$

The factors for x^2 are $x \cdot x$. The factors for 10 are (1 and 10) or (2 and 5). Let's try (1 and 10) first.

$(x \pm 1)(x \pm 10)$

Determine the signs. To get a +10 for the third term, the signs need to be $+ \cdot +$ or $- \cdot -$. Look at the sign in front of the second term. It is a negative sign, so the signs you want will be $- \cdot -$.

$(x-1)(x-10)$

Check using FOIL. $(x-1)(x-10) = x^2 - 10x - 1x + 10 = x^2 - 11x + 10$

This does not give you the original trinomial, so try the other pair of factors, 2 and 5, instead of 1 and 10.

$(x-2)(x-5) = x^2 - 5x - 2x + 10 = x^2 - 7x + 10$

This pair of factors gives you the original trinomial, so you know you have the correct answer.

Example: $x^2 - 5x + 4$

The factors of x^2 are $x \cdot x$. The factors of 4 are (1 and 4) or (2 and 2). Try (1 and 4) first.

$(x \pm 1)(x \pm 4)$

Determine the sign. To get a +4, your numbers must be either both positive or both negative. The sign in front of the second term is negative, so they will both be negative.

$(x-1)(x-4)$

Check using FOIL. $(x-1)(x-4) = x^2 - 4x - 1x + 4 = x^2 - 5x + 4$

You got the correct answer on the first try! Way to go!

Practice

Factor using the trinomial method.

_______**31.** $x^2 + 4x + 4$

_______**32.** $x^2 + 6x + 8$

_______**33.** $x^2 - 4x + 3$

_______**34.** $x^2 + 7x + 12$

_______**35.** $x^2 - 10x + 16$

_______**36.** $x^2 - 15x + 14$

_______**37.** $x^2 + 9x + 20$

_______**38.** $x^2 - 12x + 20$

_______**39.** $x^2 + 9x + 20$

_______**40.** $x^2 - 11x + 30$

▶ When the Sign in Front of the Third Term Is Negative

When the sign in front of the third term is negative, you can use similar steps to factor the trinomial. Look carefully at the following examples.

Example: $x^2 + 3x - 4$

The factors for x^2 are (x and x). The factors for 4 are (1 and 4) or (2 and 2). To get a negative number for the third term, one sign has to be positive and one sign has to be negative. Let's try the factors (1 and 4) first.

$$(x + 1)(x - 4)$$

In order to get –4 for the third term, one factor must be positive and one factor must be negative. You don't know which one is positive and which is negative, so use trial and error until you get the right answer.

$$(x + 1)(x - 4) = x^2 - 4x + 1x - 4 = x^2 - 3x - 4$$

This is not the original trinomial. You have the right numbers, but the signs aren't the same. Try changing the signs in the factors.

$$(x - 1)(x + 4) = x^2 + 4x - 1x - 4 = x^2 + 3x - 4$$

This does check out, so the correct factors for the trinomial are $(x - 1)(x + 4)$.

Example: $x^2 - x - 12$

The factors for x^2 are (x and x). The factors for 12 are (1 and 12), (2 and 6), and (3 and 4). You have –1 for the second term. When you look at the choices of factors, the factors (3 and 4) have a difference of one, so try that combination first.

The sign in front of the third term is negative, so that tells you that you will have one positive factor and one negative factor in your answer.

$$(x - 3)(x + 4) = x^2 + 4x - 3x - 12 = x^2 + x - 12$$

You have the right numbers, but the signs are different. Try changing the signs of the factors.

$$(x + 3)(x - 4) = x^2 - 4x + 3x - 12 = x^2 - x - 12$$

This checks out, so the correct factors are $(x + 3)(x - 4)$.

Tip

When the factors have one positive sign and one negative sign, subtract the combination of factors to get the coefficient of the second term.

Practice

Factor the trinomials.

______**41.** $x^2 - 4x - 21$

______**42.** $x^2 + x - 6$

______**43.** $x^2 + x - 12$

______**44.** $x^2 + 3x - 10$

______**45.** $x^2 - 3x - 10$

______**46.** $x^2 - 7x - 8$

______**47.** $x^2 + 6x - 16$

______**48.** $x^2 - 4x - 21$

______**49.** $x^2 + x - 30$

______**50.** $x^2 - 3x - 18$

Skill Building until Next Time

When you factor polynomials, recognizing perfect squares is extremely helpful. Take the time to make a list of all the numbers that are perfect squares between 0 and 400. When you have developed the list, save it so that you can use it when you are doing lessons that require factoring. It will be a time-saver! Your list will begin with 1, 4, 9, 16, …

LESSON

16 Using Factoring

LESSON SUMMARY

In this lesson, you will learn how to factor expressions that have a coefficient other than one for the first term. You will learn the types of factoring you need to use for different problems. You will also factor expressions that require more than one type of factoring.

► Factoring Trinomials That Have a Coefficient Other Than One for the First Term

Factoring trinomials that have a coefficient other than one for the first term becomes complex because it increases the choices of factors. You will use what you have already learned in previous lessons about factoring trinomials, but you may need to go through many more steps to determine the factors. Study the following examples.

Example: $2x^2 + 7x + 5$

The factors for $2x^2$ are ($2x$ and x).	$(2x \pm [\quad])(x \pm [\quad])$
The signs for the factors will both be positive.	$(2x + [\quad])(x + [\quad])$
The factors for 5 are (5 and 1).	$(2x + 5)(x + 1)$
Check using FOIL.	$2x^2 + 2x + 5x + 5$
	$= 2x^2 + 7x + 5$

The result is the original trinomial, so the factors of $2x^2 + 7x + 5 = (2x + 5)(x + 1)$.

Example: $6x^2 - 13x + 6$

The factors for $6x^2$ are ($1x$ and $6x$) and ($2x$ and $3x$). First, let's try $1x$ and $6x$.

$$(x \pm [\quad])(6x \pm [\quad])$$

The signs will both be negative. Remember, the sign in front of the third term is positive, which tells you both signs will be the same. The sign in front of the second term is negative, which tells you both signs will be negative.

$$(x - [\quad])(6x - [\quad])$$

The factors for the third term are (1 and 6) and (2 and 3). Try (1 and 6).

$$(x - 1)(6x - 6)$$

Test using FOIL.

$$6x^2 - 6x - 6x + 6$$
$$= 6x^2 - 12x + 6$$

You can see this isn't correct because the original trinomial is different. So, try interchanging the 1 and 6.

$$(x - 6)(6x - 1)$$
$$6x^2 - x - 36x + 6$$
$$= 6x^2 - 37x + 6$$

That didn't work either! Now, try using (2 and 3).

$$(x - 2)(6x - 3)$$

Test using FOIL.

$$6x^2 - 3x - 12x + 6$$
$$6x^2 - 15x + 6$$

That's not it. Try interchanging the 2 and 3.

$$(x - 3)(6x - 2)$$

Test using FOIL.

$$6x^2 - 2x - 18x + 6$$
$$= 6x^2 - 20x + 6$$

It's still not right, but KEEP TRYING!

Now try using $2x$ and $3x$ for the first term.

$$(2x - 1)(3x - 6)$$

Test using FOIL.

$$6x^2 - 12x - 3x + 6$$
$$= 6x^2 - 15x + 6$$

Interchange the 1 and 6.

$$(2x - 6)(3x - 1)$$

Test using FOIL.

$$6x^2 - 2x - 18x + 6$$
$$= 6x^2 - 20x + 6$$

Try using the factors 2 and 3.

$$(2x - 2)(3x - 3)$$

Test using FOIL.

$$6x^2 - 6x - 6x + 6$$
$$= 6x^2 - 12x + 6$$

Interchange the 2 and 3.

$$(2x - 3)(3x - 2)$$

Test using FOIL.

$$6x^2 - 4x - 9x + 6$$
$$= 6x^2 - 13x + 6$$

FINALLY! Factoring trinomials with a coefficient other than one in front of the first term can be complex, as shown in the previous example. But don't get discouraged. You could have found the correct factors on the first try. You will find that the more experience you have factoring trinomials, the easier it will become.

Example: $4x^2 + x - 5$

The factors of $4x^2$ are (x and $4x$) and ($2x$ and $2x$). Try (x and $4x$).

$(x \pm [\quad])(4x \pm [\quad])$

The factors of 5 are (1 and 5). $(x \pm 1)(4x \pm 5)$

Because there is a negative sign in front of the 5, you will have one positive sign and one negative sign.

$(x + 1)(4x - 5)$

Test using FOIL. $4x^2 - 5x + 4x - 5$

$= 4x^2 - x - 5$

You have the correct terms, but the signs aren't right. Interchange the positive and negative signs.

$(x - 1)(4x + 5)$

Test using FOIL. $4x^2 + 5x - 4x - 5$

$= 4x^2 + x - 5$

The factors of $4x^2 + x - 5 = (x - 1)(4x + 5)$.

Practice

Factor using the trinomial method.

______ **1.** $2x^2 + 7x + 3$

______ **2.** $5x^2 + 13x - 6$

______ **3.** $7x^2 + 3x - 4$

______ **4.** $4x^2 - 21x + 5$

______ **5.** $6x^2 + x - 7$

______ **6.** $8x^2 - 3x - 5$

______ **7.** $8x^2 - 6x - 5$

______ **8.** $9x^2 + 3x - 2$

______ **9.** $6x^2 - 17x + 5$

______ **10.** $5x^2 - 7x - 6$

▶ Factoring Using Any Method

When you factor algebraic expressions, you need to analyze the expression to determine which type of factoring to use. The three methods of factoring you have learned are:

1. Greatest common factor
2. Difference of two squares
3. Trinomials

Tip

When factoring an algebraic expression, always look for the greatest common factor first.

Practice

Factor using the appropriate method.

______**11.** $2x^2 + 4$

______**12.** $x^2 - 4$

______**13.** $x^2 + 8x - 33$

______**14.** $9x^2 - 25$

______**15.** $x^2 + 4x + 4$

______**16.** $10x^4 + 12x^2 - 6x$

______**17.** $121x^2 - 1$

______**18.** $x^2 + 9$

______**19.** $x^2 + 10x + 9$

______**20.** $x^2 + 2x - 15$

______**21.** $18x^2 + 27a$

______**22.** $6a^2 - 30ab$

______**23.** $49x^2 - 4$

______**24.** $c^2 - 11c + 30$

______**25.** $a^2 - b^2$

______**26.** $b^2 + 3b - 18$

______**27.** $n^2 - 2n - 35$

______**28.** $24x + 6$

______**29.** $9x^2 - 100$

______**30.** $5x^2 + 7x - 6$

______**31.** $6x^2 - 7x - 3$

______**32.** $r^2 - 5r - 24$

______**33.** $f^2 + 5f - 36$

______**34.** $3x^3y + 6x^2y^2 - 9xy^3$

______**35.** $15x^2 - 7x - 2$

______**36.** $25a^2 - 64$

______**37.** $48x^3y^3 - 18x^4y$

______**38.** $6x^2 + 25x + 11$

______**39.** $10mn + 5m^2n^3 - 20m^3n^2$

______**40.** $25x^2 + 1$

► Factoring Using More Than One Method

Sometimes it may be necessary to use more than one method of factoring on the same expression. Always check for the greatest common factor first.

Example: $4x^2 - 4$

Look for the greatest common factor first. $4(x^2 - 1)$

You aren't finished because $x^2 - 1$ is the difference of two squares.

$= 4(x - 1)(x + 1)$

Example: $2x^2 + 8x + 8$

Look for the greatest common factor. $2(x^2 + 4x + 4)$

Factor the trinomial. $2(x + 2)(x + 2)$

Practice

Factor completely.

______**41.** $3x^2 - 27$

______**42.** $4x^2 - 64$

______**43.** $2x^2 + 12x + 18$

______**44.** $2x^2 + 4x - 6$

______**45.** $3x^2 + 21x + 30$

______**46.** $4x^6 - 100$

______**47.** $27x^2 - 75y^2$

______**48.** $12x^2 - 36x - 21$

______**49.** $3x^2 - 24x + 36$

______**50.** $x^4 - 81$

Skill Building until Next Time

The next time you go to the grocery store, look for a special and calculate the amount saved if you bought a quantity of 10 items. For example, bananas normally sell for $.59 a pound. When they are on sale for 3 pounds for $1.00, how much will you save if you buy 10 pounds of bananas? The difference between the original price and the sale price is ($.59 – $.33). Multiply this amount times 10 to calculate the amount you would save when you buy 10 pounds:

10($.59 – $.33) = $2.60

Store owners can use factors like these with a variable to help them set prices to make their desired profit.

LESSON

17 ▶ Solving Quadratic Equations

LESSON SUMMARY

In this lesson, you will use factoring to solve quadratic equations. You will also use quadratic equations to solve real-life problems.

▶ What Is a Quadratic Equation?

A *quadratic equation* is an equation that does not graph into a straight line. The graph of a quadratic equation will be a smooth curve. This is different from a linear equation, which is an equation that will graph into a straight line. An equation is a quadratic equation if the highest exponent of the variable is 2.

Here are some examples of quadratic equations:

$x^2 = 4$
$x^2 + 3 = 0$
$2x^2 + 5 = 10$
$x^2 + 4x + 4 = 0$
$5x^2 - 1 = 0$

You have learned how to solve linear equations, but you can't solve quadratic equations with only this knowledge. You will understand the importance of studying factoring after you finish this lesson. Factoring is one of the methods you will use to solve quadratic equations. So read on to find out how you can master quadratic equations by using factoring.

▶ Solving Quadratic Equations Using Factoring

Why is the equation, $x^2 = 4$, a quadratic equation? It is a quadratic equation because the variable has an exponent of 2. To solve a quadratic equation, first make one side of the equation zero.

Example: $x^2 = 4$
Subtract 4 from both sides of the equation to make one side of the equation zero.

$$x^2 - 4 = 4 - 4$$

Simplify. $x^2 - 4 = 0$

The next step is to factor $x^2 - 4$. It can be factored as the difference of two squares.

$$(x - 2)(x + 2) = 0$$

If $ab = 0$, you know that either a or b or both factors have to be zero since a times $b = 0$. This is called the *zero product property,* and it says that if the product of two numbers is zero, then one or both of the numbers have to be zero. You can use this idea to help solve quadratic equations with the factoring method.

Use the *zero product property,* and set each factor equal to zero.

$$(x - 2) = 0 \text{ and } (x + 2) = 0$$

When you use the *zero product property,* you get linear equations that you already know how to solve.

Solve the equation. $x - 2 = 0$
Add 2 to both sides of the equation. $x - 2 + 2 = 0 + 2$
Simplify. $x = 2$

Solve the equation. $x + 2 = 0$
Subtract 2 from both sides of the equation. $x + 2 - 2 = 0 - 2$
Simplify. $x = -2$

You got two values for x. The two solutions for x are 2 and –2. All quadratic equations have two solutions. The exponent 2 in the equation tells you that the equation is quadratic, and it also tells you that you will have two answers.

Before you can factor an expression, the expression must be arranged in descending order. An expression is in descending order when you start with the largest exponent and descend to the smallest, as shown in this example: $2x^2 + 5x + 6 = 0$.

Tip

All quadratic equations have two solutions. The exponent of 2 in the equation tells you to expect two answers.

Example: $x^2 - 3x - 4 = 0$

Factor the trinomial $x^2 - 3x - 4$.	$(x-4)(x+1) = 0$
Set each factor equal to zero.	$x - 4 = 0$ and $x + 1 = 0$
Solve the equation.	$x - 4 = 0$
Add 4 to both sides of the equation.	$x - 4 + 4 = 0 + 4$
Simplify.	$x = 4$
Solve the equation.	$x + 1 = 0$
Subtract 1 from both sides of the equation.	$x + 1 - 1 = 0 - 1$
Simplify.	$x = -1$

The two solutions for the quadratic equation are 4 and –1.

Example: $4x^2 = 100$

Make one side of the equation zero. Subtract 100 from both sides of the equation.

	$4x^2 - 100 = 100 - 100$
Simplify.	$4x^2 - 100 = 0$
Factor out the greatest common factor.	$4(x^2 - 25) = 0$
Factor using the difference of two squares.	$4(x-5)(x+5) = 0$
Divide both sides of the equation by 4.	$\frac{4(x-5)(x+5)}{4} = \frac{0}{4}$
Simplify.	$(x-5)(x+5) = 0$
Set each factor equal to zero.	$x - 5 = 0$ and $x + 5 = 0$
Solve the equation.	$x - 5 = 0$
Add 5 to both sides of the equation.	$x - 5 + 5 = 0 + 5$
Simplify.	$x = 5$
Solve the equation.	$x + 5 = 0$
Subtract 5 from both sides of the equation.	$x + 5 - 5 = 0 - 5$
Simplify.	$x = -5$

The solutions for the quadratic equation $4x^2 = 100$ are 5 and –5.

Tip

When you have an equation in factor form, disregard any factor that is a number. For example, in this equation, $4(x - 5)(x + 5) = 0$, disregard the 4. It will have no effect on your two solutions.

Example: $2x^2 - 33 = -1$

Add 1 to both sides of the equation.	$2x^2 - 33 = -1 + 1$
Simplify both sides of the equation.	$2x^2 - 32 = 0$

Take out the common factor.	$2(x^2 - 16) = 0$
Factor the difference of two squares.	$2(x - 4)(x + 4) = 0$
Disregard the 2 and set the other factors equal to zero.	$x - 4 = 0$ and $x + 4 = 0$
Solve the equation.	$x - 4 = 0$
Add 4 to both sides of the equation.	$x - 4 + 4 = 0 + 4$
Simplify both sides of the equation.	$x = 4$
Solve the equation.	$x + 4 = 0$
Subtract 4 from both sides of the equation.	$x + 4 - 4 = 0 - 4$
Simplify both sides of the equation.	$x = -4$

The solutions are 4 and –4.

Tip

When both your solutions are the same number, this is called a *double root*. You will get a double root when both factors are the same.

Practice

Solve the quadratic equations using factoring.

______ **1.** $x^2 - 36 = 0$

______ **2.** $2x^2 - 50 = 0$

______ **3.** $x^2 = 49$

______ **4.** $4x^2 = 100$

______ **5.** $x^2 - 50 = -1$

______ **6.** $x^2 + x - 2 = 0$

______ **7.** $x^2 + 7x - 18 = 0$

______ **8.** $x^2 + 4x = 45$

______ **9.** $x^2 - 6 = 30 - 3x^2$

______ **10.** $x^2 + 11x = -24$

______ **11.** $2x^2 - 16x - 18 = 0$

______ **12.** $6x^2 - 12x + 6 = 0$

______ **13.** $x^2 - 4x = 45$

______ **14.** $2x^2 - 5x - 7 = 0$

______ **15.** $3x^2 - 20x = 7$

______ **16.** $2x^2 + 2x - 4 = 0$

______ **17.** $x^2 + 2x = 15$

______ **18.** $2x^2 + 6x - 20 = 0$

______ **19.** $16x^2 + 2x = 2x + 9$

______ **20.** $15x^2 + 84x = 36$

Applications

Solve the word problems using quadratic equations. First, look carefully at the example given.

Example: You have a patio that is 8 ft. by 10 ft. You want to increase the size of the patio to 168 square ft. by adding the same length to both sides of the patio. Let x = the length you will add to each side of the patio.

You find the area of a rectangle by multiplying the length times the width. The new area of the patio will be 168 square ft.

$(x + 8)(x + 10) = 168$

FOIL the factors $(x + 8)(x + 10)$.	$x^2 + 10x + 8x + 80 = 168$
Simplify.	$x^2 + 18x + 80 = 168$
Subtract 168 from both sides of the equation.	$x^2 + 18x + 80 - 168 = 168 - 168$
Simplify both sides of the equation.	$x^2 + 18x - 88 = 0$
Factor.	$(x + 22)(x - 4) = 0$
Set each factor equal to zero.	$x + 22 = 0$ and $x - 4 = 0$
Solve the equation.	$x + 22 = 0$
Subtract 22 from both sides of the equation.	$x + 22 - 22 = 0 - 22$
Simplify both sides of the equation.	$x = -22$
Solve the equation.	$x - 4 = 0$
Add 4 to both sides of the equation.	$x - 4 + 4 = 0 + 4$
Simplify both sides of the equation.	$x = 4$

Because this is a quadratic equation, you can expect two answers. The answers are 4 and –22. However, –22 is not a reasonable answer. You cannot have a negative length. Therefore, the only answer is 4.

Check: The original dimensions of the patio were 8 ft. by 10 ft. If you were to add 4 to each side, the new dimensions would be 12 ft. by 14 ft. When you multiply 12 times 14, you get 168 square ft., which is the new area you wanted.

Practice

21. A rectangle is 6 feet wide and 4 feet tall. What amount should be added to the height and width to make the area 63 square feet?

22. The Jensens are installing a swimming pool in their backyard. The dimensions of the pool are 25 ft. by 30 ft. They want to have a sidewalk around the pool, but the area of the pool and the sidewalk must be 1,050 square ft. What is the length that should be added to each side of the pool to get an area of 1,050 square ft? What will be the width of the sidewalk?

23. Jessica is making a quilt. Each quilting square is 4 inches by 4 inches. She wants to put a border around each square so that the new area of the quilting square will be 36 square inches. What length does she have to add to each side? What is the width of the border?

24. A dairy is required to have a waste holding pond for runoff. Mr. Van Beek plans to build a pond that will be 20 ft by 15 ft. There will be a border of equal length around the pond. The total area of the pond and the border is 500 square ft. How wide will the border be?

25. Mr. Ingram wants to add a garage to his home. The dimensions of his home are 50 ft by 20 ft. When securing his building permit, he found that his home with the garage cannot be more than 1,400 square ft. How long can he extend the side of his house that is 50 ft. long so that his remodeling project will follow the code?

Skill Building until Next Time

Find your favorite snapshot. Frame the picture with a border. Then, determine the size frame that you will need to fit the snapshot and the border. How wide will you make the border? What considerations did you need to make when framing the picture?

LESSON

18 Simplifying Radicals

LESSON SUMMARY

This lesson defines radicals and shows you how to simplify them. You will also learn how to add, subtract, multiply, and divide radicals.

What Is a Radical?

You have seen how the addition in $x + 5 = 11$ can be undone by subtracting 5 from both sides of the equation. You have also seen how the multiplication in $3x = 21$ can be undone by dividing both sides by 3. Taking the *square root* (also called a *radical*) is the way to undo the exponent from an equation like $x^2 = 25$.

The exponent in 7^2 tells you to square 7. You multiply $7 \cdot 7$ and get $7^2 = 49$.

The *radical sign* $\sqrt{\ }$ in $\sqrt{36}$ tells you to find the positive number whose square is 36. In other words, $\sqrt{36}$ asks: What number times itself is 36? The answer is $\sqrt{36} = 6$ because $6 \cdot 6 = 36$.

The number inside the radical sign is called the *radicand*. For example, in $\sqrt{9}$, the radicand is 9.

► Square Roots of Perfect Squares

The easiest radicands to deal with are perfect squares. Since they appear so often, it is useful to learn to recognize the first few perfect squares: $0^2 = 0$, $1^2 = 1$, $2^2 = 4$, $3^2 = 9$, $4^2 = 16$, $5^2 = 25$, $6^2 = 36$, $7^2 = 49$, $8^2 = 64$, $9^2 = 81$, $10^2 = 100$, $11^2 = 121$, and $12^2 = 144$. It is even easier to recognize when a variable is a perfect square because the exponent is even. For example: $x^{14} = x^7 \cdot x^7$ and $a^8 = a^4 \cdot a^4$.

Example: $\sqrt{64x^2y^{10}}$

Write as a square. $\sqrt{8xy^5 \cdot 8xy^5}$

Evaluate. $8xy^5$

You could also have split the radical into parts and evaluated them separately:

Example: $\sqrt{64x^2y^{10}}$

Split into perfect squares. $\sqrt{64 \cdot x^2 \cdot y^{10}}$

Write as squares. $\sqrt{8 \cdot 8} \cdot \sqrt{x \cdot x} \cdot \sqrt{y^5 \cdot y^5}$

Evaluate. $8 \cdot x \cdot y$

Multiply together. $8xy$

If your radical has a coefficient like $3\sqrt{25}$, evaluate the square root before multiplying: $3\sqrt{25} = 3 \cdot 5 = 15$.

Practice

Simplify the radicals.

_____ **1.** $\sqrt{49}$

_____ **2.** $\sqrt{81}$

_____ **3.** $\sqrt{144}$

_____ **4.** $-\sqrt{64}$

_____ **5.** $4\sqrt{4}$

_____ **6.** $-2\sqrt{9}$

_____ **7.** $\sqrt{a^2}$

_____ **8.** $5\sqrt{36}$

_____ **9.** $\sqrt{1{,}600}$

_____ **10.** $8\sqrt{0}$

_____ **11.** $\sqrt{0.04}$

_____ **12.** $\sqrt{100x^4}$

_____ **13.** $-2\sqrt{4a^8}$

_____ **14.** $3\sqrt{25x^2y^{18}}$

_____ **15.** $-4\sqrt{400a^6b^2}$

_____ **16.** $5\sqrt{144x^4y^2}$

▶ Simplifying Radicals

Not all radicands are perfect squares. There is no whole number that, multiplied by itself, equals 5. With a calculator, you can get a decimal that squares to very close to 5, but it won't come out exactly. The only precise way to represent the square root of five is to write $\sqrt{5}$. It cannot be simplified any further.

There are three rules for knowing when a radical cannot be simplified any further:

1. The radicand contains no factor, other than one, that is a perfect square.
2. The radicand cannot be a fraction.
3. The radical cannot be in the denominator of a fraction.

When the Radicand Contains a Factor That Is a Perfect Square

To determine if a radicand contains any factors that are perfect squares, factor the radicand completely. All the factors must be prime. A number is prime if its only factors are 1 and the number itself. A prime number cannot be factored any further.

For example, here's how you simplify $\sqrt{12}$. The number 12 can be factored into $12 = 2 \cdot 6$. This is not completely factored because 6 is not prime. The number 6 can be further factored $6 = 2 \cdot 3$. The number 12 completely factored is $2 \cdot 2 \cdot 3$.

The radical $\sqrt{12}$ can be written as $\sqrt{2 \cdot 2 \cdot 3}$. This can be split up into $\sqrt{2 \cdot 2} \cdot \sqrt{3}$. Since $\sqrt{2 \cdot 2} = 2$, the simplified form of $\sqrt{12}$ is $2\sqrt{3}$.

Example: $\sqrt{18}$

Factor completely. $\sqrt{2 \cdot 3 \cdot 3}$

Separate out the perfect square $3 \cdot 3$. $\sqrt{3 \cdot 3} \cdot \sqrt{2}$

Simplify. $3\sqrt{2}$

Example: $\sqrt{60}$

Factor completely. $\sqrt{6 \cdot 10}$

Neither 6 nor 10 is prime. Both can be factored further. $\sqrt{2 \cdot 3 \cdot 2 \cdot 5}$

Separate out the perfect square $2 \cdot 2$. $\sqrt{2 \cdot 2} \cdot \sqrt{3 \cdot 5}$

Because $3 \cdot 5$ contains no perfect squares, it cannot be simplified further. $2\sqrt{15}$

Example: $\sqrt{32}$

Factor completely. $\sqrt{2 \cdot 16}$

The number 16 is not prime. It can be factored. $\sqrt{2 \cdot 2 \cdot 8}$

The number 8 is not prime. It can be factored. $\sqrt{2 \cdot 2 \cdot 2 \cdot 4}$

The number 4 is not prime. It can be factored. $\sqrt{2 \cdot 2 \cdot 2 \cdot 2 \cdot 2}$

You have two sets of perfect squares, $2 \cdot 2$ and $2 \cdot 2$. The square root of each is 2, so you have two square roots of 2. The square roots go outside the radical. You then multiply the numbers that are outside the radical.

$2 \cdot 2\sqrt{2}$

Simplify. The product of 2 times 2 gives you 4. $4\sqrt{2}$

Shortcut: You may have noticed that in the step $\sqrt{2 \cdot 16}$ that 16 is a perfect square, and the square root of 16 is 4. This would have given you the answer $4\sqrt{2}$. This is a shortcut you can use. Use the shorter method whenever you see one.

Example: $\sqrt{50x^3}$

Factor completely.	$\sqrt{2 \cdot 5 \cdot 5 \cdot x \cdot x \cdot x}$
Separate the perfect square $5 \cdot 5$ and $x \cdot x$	$\sqrt{5 \cdot 5} \cdot \sqrt{x \cdot x} \cdot \sqrt{2 \cdot x}$
Simplify.	$5x\sqrt{2x}$

Example: $\sqrt{9x^2y^3}$

Rewrite the radicand as the product of perfect squares.	$\sqrt{9 \cdot x^2 \cdot y^2 \cdot y}$
Take out the square roots.	$3xy\sqrt{y}$

Practice

Simplify the radicals.

______**17.** $\sqrt{8}$

______**18.** $\sqrt{20}$

______**19.** $\sqrt{54}$

______**20.** $\sqrt{40}$

______**21.** $\sqrt{72}$

______**22.** $\sqrt{27}$

______**23.** $\sqrt{28}$

______**24.** $\sqrt{160}$

______**25.** $\sqrt{200}$

______**26.** $\sqrt{44}$

______**27.** $\sqrt{225}$

______**28.** $\sqrt{500}$

______**29.** $\sqrt{1{,}200}$

______**30.** $\sqrt{11}$

______**31.** $\sqrt{3x^2y^2}$

______**32.** $\sqrt{4b^6}$

______**33.** $\sqrt{8c^4d}$

______**34.** $\sqrt{80a^2b^3c^4}$

______**35.** $\sqrt{20a^5b^6c}$

______**36.** $\sqrt{500d^{13}}$

When the Radicand Contains a Fraction

The radicand cannot be a fraction. If you get rid of the denominator in the radicand, then you no longer have a fraction. This process is called rationalizing the denominator. Your strategy will be to make the denominator a perfect square. To do that, you multiply the denominator by itself. However, if you multiply the denominator of a fraction by a number, you must multiply the numerator of the fraction by the same number. Take a look at the following examples.

Example: $\sqrt{\frac{1}{2}}$

Make the denominator a perfect square. $\sqrt{\frac{1}{2} \cdot \frac{2}{2}}$

Take out the square roots. One is a perfect square and so is $2 \cdot 2$. $\frac{1}{2}\sqrt{2}$

Example: $\sqrt{\frac{2}{3}}$

Make the denominator a perfect square. $\sqrt{\frac{2}{3} \cdot \frac{3}{3}}$

The number 1 is considered a factor of all numbers. If the numerator does not contain a perfect square, then 1 will be the perfect square and will be in the numerator. Take the square root of 1 in the numerator and $3 \cdot 3$ in the denominator. The product of $2 \cdot 3$ will give you 6 for the radicand.

$\frac{1}{3}\sqrt{6}$

Example: $\sqrt{\frac{3x}{2}}$

Make the denominator a perfect square. $\sqrt{\frac{3x}{2} \cdot \frac{2}{2}}$

Take the square roots. $\frac{1}{2}\sqrt{6x}$

Practice

Simplify the radicals.

_______**37.** $\sqrt{\frac{2}{5}}$

_______**38.** $\sqrt{\frac{2x^2}{3}}$

_______**39.** $\sqrt{\frac{a^2b^2}{2}}$

_______**40.** $\sqrt{\frac{2}{7x}}$

_______**41.** $\sqrt{\frac{5x^3}{3}}$

_______**42.** $\sqrt{\frac{20}{11}}$

When There Is a Radical in the Denominator

When you have a radical in the denominator, the expression is not in simplest form. The expression $\frac{2}{\sqrt{3}}$ contains a radical in the denominator. To get rid of the radical in the denominator, rationalize the denominator. In other words, make the denominator a perfect square. To do that, you need to multiply the denominator by itself.

Example: $\frac{2}{\sqrt{3}} \cdot \frac{\sqrt{3}}{\sqrt{3}}$

Simplify. $\frac{2\sqrt{3}}{\sqrt{9}}$

The number 9 is a perfect square. $\frac{2\sqrt{3}}{3}$

Example: $\frac{5}{\sqrt{2}}$

Rationalize the denominator. $\frac{5}{\sqrt{2}} \cdot \frac{\sqrt{2}}{\sqrt{2}}$

Simplify. $\frac{5\sqrt{2}}{\sqrt{4}}$

Take the square root of 4. $\frac{5\sqrt{2}}{2}$

Example: $\frac{\sqrt{6}}{\sqrt{2}}$

Rationalize the denominator. $\frac{\sqrt{6}}{\sqrt{2}} \cdot \frac{\sqrt{2}}{\sqrt{2}}$

Simplify. $\frac{\sqrt{12}}{\sqrt{4}}$

You aren't finished yet because both radicands contain perfect squares.

$\frac{\sqrt{3 \cdot 4}}{\sqrt{4}}$

Take the square root of 4. $\frac{2\sqrt{3}}{2}$

Finished? Not quite. You can divide 2 into 2, or cancel the 2s. $\sqrt{3}$

Practice

Simplify the radicals.

43. $\frac{3}{\sqrt{7}}$

44. $\frac{6}{\sqrt{5}}$

45. $\frac{\sqrt{2}}{\sqrt{3}}$

46. $\frac{2\sqrt{3}}{\sqrt{6}}$

47. $\frac{5}{\sqrt{x}}$

48. $\frac{3}{\sqrt{2y}}$

49. $\frac{7\sqrt{2}}{\sqrt{7}}$

50. $\frac{\sqrt{8ab}}{\sqrt{b}}$

► Adding and Subtracting Radicals

It is easy to add and subtract radicals. You can add and subtract radicals if the radicands are the same. For example, you can add $3\sqrt{2}$ and $5\sqrt{2}$ because the radicands are the same. To add or subtract radicals, you add the number in front of the radicals and leave the radicand the same. When you add $3\sqrt{2} + 5\sqrt{2}$, you add the 3 and the 5, but the radicand $\sqrt{2}$ stays the same. The answer is $8\sqrt{2}$.

Tip

You can add or subtract radicals *only* when the radicand is the same. You add radicals by adding the number in front of the radicals and keeping the radicand the same. When you subtract radicals, you subtract the numbers in front of the radicals and keep the radicand the same.

Example: $2\sqrt{5} + 7\sqrt{5}$

Add the numbers in front of the radicals. $9\sqrt{5}$

Example: $11\sqrt{5} - 4\sqrt{5}$

Subtract the numbers in front of the radicals. $7\sqrt{5}$

Example: $4\sqrt{3} + 2\sqrt{5} + 6\sqrt{3}$

You can only add the radicals that are the same. $10\sqrt{3} + 2\sqrt{5}$

Example: $5\sqrt{8} + 6\sqrt{8}$

Add the radicals. $11\sqrt{8}$

But $\sqrt{8}$ contains a factor that is a perfect square, so you aren't finished because your answer is not in simplest form. $11\sqrt{2 \cdot 4}$

Take out the square root of 4. $2 \cdot 11\sqrt{2}$

Simplify. $22\sqrt{2}$

Practice

Add and subtract the radicals. Simplify the radical when necessary.

______ **51.** $3\sqrt{7} + 8\sqrt{7}$

______ **52.** $11\sqrt{3} - 8\sqrt{3}$

______ **53.** $5\sqrt{2} + 6\sqrt{2} - 3\sqrt{2}$

______ **54.** $3\sqrt{6} + 2\sqrt{2} - 5\sqrt{6}$

______ **55.** $4\sqrt{a} + 13\sqrt{a}$

______ **56.** $\sqrt{75} - \sqrt{20} + 3\sqrt{5}$

______ **57.** $9\sqrt{x} - 4\sqrt{y} + 2\sqrt{x}$

______ **58.** $\sqrt{12} + 3\sqrt{3}$

______ **59.** $2\sqrt{18} + 6\sqrt{2}$

______ **60.** $3\sqrt{5} + \sqrt{20} - \sqrt{7}$

► Multiplying and Dividing Radicals

It is easy to multiply and divide radicals. To multiply radicals like $4\sqrt{3}$ times $2\sqrt{2}$, you multiply the numbers in front of the radicals: 4 times 2. Then multiply the radicands: 3 times 2. The answer is $8\sqrt{6}$.

Example: $5\sqrt{3} \cdot 2\sqrt{2}$

Multiply the numbers in front of the radicals. Then multiply the radicands.

$10\sqrt{6}$

Example: $2\sqrt{6} \cdot 3\sqrt{3}$

Multiply the numbers in front of the radicals. Then multiply the radicands.

$6\sqrt{18}$

However, you are not finished yet because 18 contains the factor 9,
which is a perfect square. $6\sqrt{2 \cdot 9}$

Take out the square root of 9. $3 \cdot 6\sqrt{2}$

Simplify. $18\sqrt{2}$

To divide the radical $4\sqrt{6}$ by $2\sqrt{3}$, divide the numbers in front of the radicals. Then divide the radicands. The answer is $2\sqrt{2}$.

Tip

When you multiply or divide radicals, the radicands do not have to be the same.

Example: $\frac{10\sqrt{6}}{5\sqrt{2}}$

Divide the numbers in front of the radical. Then divide the radicands.

$2\sqrt{3}$

Example: $\frac{8\sqrt{20}}{4\sqrt{5}}$

Divide the numbers in front of the radicals. Then divide the radicands.

$2\sqrt{4}$

However, you aren't finished yet because 4 is a perfect square.

Take the square root of 4. $2 \cdot 2$

Simplify. 4

Tip

If there is no number in front of the radical sign, it is assumed to be 1.

Practice

Multiply and divide the radicals. Simplify the radicals when necessary.

61. $7\sqrt{3} \cdot 5\sqrt{2}$

62. $\frac{14\sqrt{6}}{7\sqrt{2}}$

63. $-3\sqrt{5} \cdot 4\sqrt{2}$

64. $\frac{24\sqrt{10}}{12\sqrt{2}}$

65. $3\sqrt{a} \cdot 4\sqrt{b}$

66. $\frac{8\sqrt{x^2}}{4\sqrt{x}}$

67. $\frac{6\sqrt{20}}{2\sqrt{5}}$

68. $3\sqrt{10} \cdot 5\sqrt{10}$

69. $4\sqrt{6} \cdot 5\sqrt{3}$

70. $\frac{30\sqrt{15}}{4\sqrt{10}}$

Skill Building until Next Time

Think about the formula for the area of a rectangle: $A = lw$. Why do you use "square" when you multiply a number by itself? What word would you use when multiplying a number by itself three times? Why isn't there a nice word for multiplying a number by itself four times?

LESSON

19 Solving Radical Equations

LESSON SUMMARY

In this lesson, you will solve equations that contain radicals. You will use the skills you have acquired throughout this book to solve these equations. You may also need to simplify radicals to solve the equations in this lesson.

► Solving Basic Radical Equations

What is a radical equation? An equation is not considered a *radical equation* unless the radicand contains a variable, like $\sqrt{x} = 3$. You know that squaring something is the opposite of taking the square root. To solve a radical equation, you square both sides.

Example: $\sqrt{x} = 3$

Square both sides.	$(\sqrt{x})^2 = 3^2$
Simplify.	$\sqrt{x} \cdot \sqrt{x} = 3 \cdot 3$
Multiply.	$\sqrt{x \cdot x} = 9$
Take the square root of x^2.	$x = 9$

To solve an equation like $x^2 = 25$ requires a little extra thought. Plug in $x = 5$ and you see that $5^2 = 25$. This means that $x = 5$ is a solution to $x^2 = 25$. However, if you plug in $x = -5$, you see that $(-5)^2 = (-5) \cdot (-5) = 25$ also. This means that $x = -5$ is also a solution to $x^2 = 25$. The equation $x^2 = 25$ has two solutions: $x = 5$ and $x = -5$. This happens so often that there is a special symbol $\pm$ that means *plus or minus*. You say that $x = \pm 5$ is the solution to $x^2 = 25$.

Remember that every quadratic equation has two solutions.

Example: $x^2 = 24$

Take the square root of both sides.	$\sqrt{x^2} = \sqrt{24}$
The answer could be + or –.	$x = \pm\sqrt{24}$
Factor out perfect squares.	$x = \pm\sqrt{4 \cdot 6}$
Simplify.	$x = \pm 2\sqrt{6}$

Practice

Solve the radical equations. Show your steps.

______ **1.** $x^2 = 81$

______ **2.** $x^2 = 50$

______ **3.** $\sqrt{x} = 8$

______ **4.** $\sqrt{n} = 3 \cdot 2$

▶ Solving Complex Radical Equations

Now that you know what a radical equation is, use what you have learned about solving equations to solve radical equations that require more than one step.

> **Tip**
>
> Before squaring both sides of an equation, get the radical on a side by itself.

Example: $\sqrt{x} + 1 = 5$

Subtract 1 from both sides of the equation.	$\sqrt{x} + 1 - 1 = 5 - 1$
Simplify.	$\sqrt{x} = 4$
Square both sides of the equation.	$(\sqrt{x})^2 = 4^2$
Simplify.	$x = 16$

Example: $3\sqrt{x} = 15$

Divide both sides of the equation by 3.	$\frac{3\sqrt{x}}{3} = \frac{15}{3}$
Simplify.	$\sqrt{x} = 5$
Square both sides of the equation.	$(\sqrt{x})^2 = 5^2$
Simplify.	$x = 25$

Example: $2\sqrt{x + 2} = 10$

Divide both sides of the equation by 2.	$\frac{2\sqrt{x+2}}{2} = \frac{10}{2}$
Simplify.	$\sqrt{x + 2} = 5$
Square both sides of the equation.	$(\sqrt{x + 2})^2 = 5^2$
Simplify.	$x + 2 = 25$
Subtract 2 from both sides of the equation.	$x + 2 - 2 = 25 - 2$
Simplify.	$x = 23$

Example: $2\sqrt{x} + 2 = 18$

Subtract 2 from both sides of the equation.	$2\sqrt{x} + 2 - 2 = 18 - 2$
Simplify.	$2\sqrt{x} = 16$
Divide both sides of the equation by 2.	$\frac{2\sqrt{x}}{2} = \frac{16}{2}$
Simplify.	$\sqrt{x} = 8$
Square both sides of the equation.	$(\sqrt{x})^2 = 8^2$
Simplify.	$x = 64$

Tip

When you multiply a radical by itself, the radical sign disappears.

Example: $\sqrt{x} \cdot \sqrt{x} = x$ and $\sqrt{x+3} \cdot \sqrt{x+3} = x+3$

Practice

Solve the radical equations. Simplify the radical when necessary. Show all your steps.

______ **5.** $\sqrt{x} + 5 = 7$

______ **6.** $\sqrt{a} - 6 = 5$

______ **7.** $4\sqrt{x} = 20$

______ **8.** $5\sqrt{w} = 10$

______ **9.** $3\sqrt{x} + 2 = 8$

______ **10.** $4\sqrt{x} - 3 = 17$

______ **11.** $\sqrt{a + 3} = 7$

______ **12.** $\sqrt{x - 5} + 2 = 11$

______ **13.** $3\sqrt{x + 2} = 21$

______ **14.** $2\sqrt{a + 3} = 6$

______ **15.** $9 - 4\sqrt{y} = -11$

______ **16.** $3 + 6\sqrt{c} = 27$

______ **17.** $\sqrt{3x + 2} = 1$

______ **18.** $\sqrt{x} = \frac{1}{9}$

______ **19.** $\sqrt{a} = 2\sqrt{3}$

______ **20.** $5\sqrt{2x + 1} + 2 = 17$

Skill Building until Next Time

The higher you climb, the farther you can see. There is a formula for this: $V = 3.5\sqrt{h}$, where h is your height, in meters, above the ground, and V is the distance, in kilometers, that you can see.

The next time you are in a building with more than one story, look out a window on different floors of the building. Do you notice a difference in how far you can see? Calculate how far the distance would be. Use 8 meters for each floor in the building. So if you were on the ninth floor, the height would be $9 \cdot 8 = 72$ meters.

LESSON

20 Using the Quadratic Formula

LESSON SUMMARY

You have solved linear equations, radical equations, and quadratic equations using the factoring method. In this lesson, you will solve quadratic equations using the quadratic formula.

▶ What Is a Quadratic Equation?

You are almost finished! After this final lesson, you will have completed all the lessons in this book.

In Lesson 17, you solved quadratic equations using factoring. What is a quadratic equation? To refresh your memory, a *quadratic equation* is an equation whose highest exponent of the variable is two. You might be asking yourself, "Why do I need to learn another method for solving quadratic equations when I already know how to solve them by using factoring?" Well, not all quadratic equations can be solved using factoring. You use the factoring method because it is faster and easier, but it will not always work. However, the quadratic formula, which is the method you will be using in this lesson, will always work.

A quadratic equation can be written in the form: $ax^2 + bx + c = 0$. The a represents the number in front of the x^2 variable. The b represents the number in front of the x variable and c is the number. For instance, in the equation $2x^2 + 3x + 5 = 0$, the a is 2, the b is 3, and the c is 5. In the equation $4x^2 - 6x + 7 = 0$, the a is 4, the b is -6, and the c is 7. In the equation $5x^2 + 7 = 0$, the a is 5, the b is 0, and the c is 7. In the equation $8x^2 - 3x = 0$, the a is 8, the b is -3, and the c is 0. Is the equation $2x + 7 = 0$ a quadratic equation? No! The equation does not contain a variable with an exponent of two. Therefore, it is not a quadratic equation.

To use the quadratic formula, you need to know the a, b, and c of the equation. However, before you can determine what a, b, and c are, the equation must be in $ax^2 + bx + c = 0$ form. The equation $5x^2 + 2x = 9$ must be transformed to $ax^2 + bx + c = 0$ form.

Example: $5x^2 + 2x = 9$

Subtract 9 from both sides of the equation. $5x^2 + 2x - 9 = 9 - 9$

Simplify. $5x^2 + 2x - 9 = 0$

In this equation, a is 5, b is 2, and c is –9.

Practice

Find a, b, and c in the following quadratic equations.

______ **1.** $4x^2 + 8x + 1 = 0$

______ **2.** $x^2 - 4x + 10 = 0$

______ **3.** $2x^2 + 3x = 0$

______ **4.** $6x^2 - 8 = 0$

______ **5.** $4x^2 = 7$

______ **6.** $3x^2 = 0$

______ **7.** $2x^2 = -3x + 4$

______ **8.** $9x^2 + 2 = 7x$

▶ What Is the Quadratic Formula?

The quadratic formula is a formula that allows you to solve any quadratic equation—no matter how simple or difficult. If the equation is written $ax^2 + bx + c = 0$, then the two solutions for x will be $x = \frac{-b \pm \sqrt{b^2 - 4ac}}{2a}$. It is the ± in the formula that gives us the two answers: one with + in that spot, and one with –. The formula contains a radical, which is one of the reasons you studied radicals in the previous lesson. To use the formula, you substitute the values of a, b, and c into the formula and then carry out the calculations.

Example: $3x^2 - x - 2 = 0$

Determine a, b, and c. $a = 3, b = -1, \text{ and } c = -2$

Take the quadratic formula. $\frac{-b \pm \sqrt{b^2 - 4ac}}{2a}$

Substitute in the values of a, b, and c. $\frac{-{}^{-}1 \pm \sqrt{(-1)^2 - 4 \cdot 3 \cdot -2}}{2 \cdot 3}$

Simplify. $\frac{1 \pm \sqrt{1 - {}^{-}24}}{6}$

Simplify more. $\frac{1 \pm \sqrt{25}}{6}$

Take the square root of 25. $\frac{1 \pm 5}{6}$

The solutions are 1 and $\frac{-2}{3}$. $\frac{1+5}{6} = \frac{6}{6} = 1$ and $\frac{1-5}{6} = \frac{-4}{6} = -\frac{2}{3}$

Practice

Solve the equations using the quadratic formula.

______ **9.** $4x^2 + 8x = 0$

______ **10.** $3x^2 = 12x$

______ **11.** $x^2 - 25 = 0$

______ **12.** $x^2 + 4x = 5$

______ **13.** $x^2 + 4x - 21 = 0$

______ **14.** $x^2 + 11x + 30 = 0$

______ **15.** $2x^2 + 5x + 3 = 0$

______ **16.** $6x^2 + 1 = 5x$

► Solving Quadratic Equations That Have a Radical in the Answer

Some equations will have radicals in their answer. The strategy for solving these equations is the same as the equations you just completed. Take a look at the following example.

Example: $3m^2 - 3m = 1$

Subtract 1 from both sides of the equation. $3m^2 - 3m - 1 = 1 - 1$

Simplify. $3m^2 - 3m - 1 = 0$

Use the quadratic formula with $a = 3$, $b = -3$, and $c = -1$. $\frac{-b \pm \sqrt{b^2 - 4ac}}{2a}$

Substitute the values for a, b, and c. $\frac{--3 \pm \sqrt{(-3)^2 - 4 \cdot 3 \cdot -1}}{2 \cdot 3}$

Simplify. $\frac{{}^{\pm}3\sqrt{9 - -12}}{6}$

Simplify. $\frac{3 \pm \sqrt{21}}{6}$

The solution to the equation is $m = \frac{3 \pm \sqrt{21}}{6}$ because one answer is $m = \frac{3 + \sqrt{21}}{6}$ and the other answer is $m = \frac{3 - \sqrt{21}}{6}$.

Practice

Solve the equations using the quadratic formula. Leave your answers in radical form.

______ **17.** $x^2 - 3x + 1 = 0$

______ **18.** $2x^2 - 2x = 5$

______ **19.** $y^2 - 5y + 2 = 0$

______ **20.** $r^2 - 7r - 3 = 0$

______ **21.** $x^2 - 3x - 5 = 0$

______ **22.** $4x^2 - 5x - 1 = 0$

______ **23.** $m^2 + 11m - 1 = 0$

______ **24.** $x^2 + 5x = 3$

Application

Solve the problem using the quadratic formula.

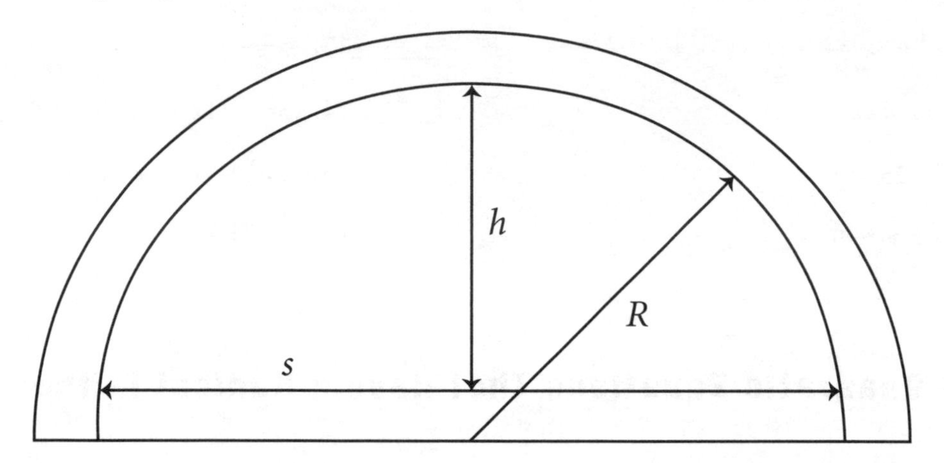

The formula $s^2 = 8Rh - 4h^2$ gives the relationship between the center height (h) of the arch, the radius of curvature (R) of the arch, and the distance (s) spanned by the arch.

Example: Find the height of an archway that has a radius of 30 ft. and spans 40 ft.

	$s^2 = 8Rh - 4h^2$
Substitute 30 for R and 40 for s.	$40^2 = 8 \cdot 30h - 4h^2$
Simplify.	$1{,}600 = 240h - 4h^2$
Add $4h^2$ to both sides of the equation.	$4h^2 + 1{,}600 = 240h - 4h^2 + 4h^2$
Simplify.	$4h^2 + 1{,}600 = 240h$
Subtract $240h$ from both sides of the equation.	$4h^2 - 240h + 1{,}600 = 240h - 240h$
Simplify.	$4h^2 - 240h + 1{,}600 = 0$
Make sure the equation is arranged in descending order.	
Use the quadratic formula to solve the equation.	
	$\frac{-b \pm \sqrt{b^2 - 4ac}}{2a}$
Substitute the values for a, b, and c.	$\frac{-(-240) \pm \sqrt{(-240)2 - 4 \cdot 4 \cdot 1{,}600}}{2 \cdot 4}$
Simplify.	$\frac{240 \pm \sqrt{57{,}600 - 25{,}600}}{8}$
Simplify.	$\frac{240 \pm \sqrt{32{,}000}}{8}$
Simplify.	$\frac{240 \pm 178.89}{8}$
Simplify.	$\frac{240 + 178.89}{8} = \frac{418.89}{8} = 52.36$
	$\frac{240 - 178.89}{8} = \frac{61.11}{8} = 7.64$

The height of 7.64 makes the better arch. Can you see why?

25. Find the height of an archway when the span is 160 ft. and the radius is 100 ft.

Skill Building until Next Time

Altitude affects the boiling point of water. If you are above sea level, the boiling point will be lower than 212° F. If you are below sea level, the boiling point of water will be higher than 212° F. The following quadratic equation approximates the number of degrees Fahrenheit below 212° (the drop *D*) at which water will boil when the altitude is above sea level: $D^2 + 520D = h$.

If you have a candy thermometer, check the temperature of boiling water. If the temperature isn't 212°, determine whether the altitude you are is above or below sea level based on the temperature of the candy thermometer in boiling water.

If you don't have a candy thermometer, you can check the effect altitude has when you bake a cake. The next time you go to a grocery store, check the directions on a box of cake mix. What adjustments do you have to make for the altitude?

Posttest

If you have completed all 20 lessons in this book, then you are ready to take the posttest to measure your progress. The posttest has 50 multiple-choice questions covering the topics you studied in this book. While the format of the posttest is similar to that of the pretest, the questions are different.

Take as much time as you need to complete the posttest. When you are finished, check your answers with the answer key at the end of the book. Along with each answer is a number that tells you which lesson of this book teaches you about the algebra skills needed for that question. Once you know your score on the posttest, compare the results with the pretest. If you scored better on the posttest than you did on the pretest, congratulations! You have profited from your hard work. At this point, you should look at the questions you missed, if any. Do you know why you missed the question, or do you need to go back to the lesson and review the concept?

If your score on the posttest doesn't show much improvement, take a second look at the questions you missed. Did you miss a question because of an error you made? If you can figure out why you missed the problem, then you understand the concept and just need to concentrate more on accuracy when taking a test. If you missed a question because you did not know how to work the problem, go back to the lesson and spend more time working that type of problem. Remember, musicians never stop practicing to perfect their skills. Take the time to understand basic algebra thoroughly. You need a solid foundation in basic algebra if you plan to use this information or progress to a higher level of algebra. Whatever your score on this posttest, keep this book for review and future reference.

LEARNINGEXPRESS ANSWER SHEET

1.	ⓐ	ⓑ	ⓒ	ⓓ
2.	ⓐ	ⓑ	ⓒ	ⓓ
3.	ⓐ	ⓑ	ⓒ	ⓓ
4.	ⓐ	ⓑ	ⓒ	ⓓ
5.	ⓐ	ⓑ	ⓒ	ⓓ
6.	ⓐ	ⓑ	ⓒ	ⓓ
7.	ⓐ	ⓑ	ⓒ	ⓓ
8.	ⓐ	ⓑ	ⓒ	ⓓ
9.	ⓐ	ⓑ	ⓒ	ⓓ
10.	ⓐ	ⓑ	ⓒ	ⓓ
11.	ⓐ	ⓑ	ⓒ	ⓓ
12.	ⓐ	ⓑ	ⓒ	ⓓ
13.	ⓐ	ⓑ	ⓒ	ⓓ
14.	ⓐ	ⓑ	ⓒ	ⓓ
15.	ⓐ	ⓑ	ⓒ	ⓓ
16.	ⓐ	ⓑ	ⓒ	ⓓ
17.	ⓐ	ⓑ	ⓒ	ⓓ
18.	ⓐ	ⓑ	ⓒ	ⓓ
19.	ⓐ	ⓑ	ⓒ	ⓓ
20.	ⓐ	ⓑ	ⓒ	ⓓ
21.	ⓐ	ⓑ	ⓒ	ⓓ
22.	ⓐ	ⓑ	ⓒ	ⓓ
23.	ⓐ	ⓑ	ⓒ	ⓓ
24.	ⓐ	ⓑ	ⓒ	ⓓ
25.	ⓐ	ⓑ	ⓒ	ⓓ
26.	ⓐ	ⓑ	ⓒ	ⓓ
27.	ⓐ	ⓑ	ⓒ	ⓓ
28.	ⓐ	ⓑ	ⓒ	ⓓ
29.	ⓐ	ⓑ	ⓒ	ⓓ
30.	ⓐ	ⓑ	ⓒ	ⓓ
31.	ⓐ	ⓑ	ⓒ	ⓓ
32.	ⓐ	ⓑ	ⓒ	ⓓ
33.	ⓐ	ⓑ	ⓒ	ⓓ
34.	ⓐ	ⓑ	ⓒ	ⓓ
35.	ⓐ	ⓑ	ⓒ	ⓓ
36.	ⓐ	ⓑ	ⓒ	ⓓ
37.	ⓐ	ⓑ	ⓒ	ⓓ
38.	ⓐ	ⓑ	ⓒ	ⓓ
39.	ⓐ	ⓑ	ⓒ	ⓓ
40.	ⓐ	ⓑ	ⓒ	ⓓ
41.	ⓐ	ⓑ	ⓒ	ⓓ
42.	ⓐ	ⓑ	ⓒ	ⓓ
43.	ⓐ	ⓑ	ⓒ	ⓓ
44.	ⓐ	ⓑ	ⓒ	ⓓ
45.	ⓐ	ⓑ	ⓒ	ⓓ
46.	ⓐ	ⓑ	ⓒ	ⓓ
47.	ⓐ	ⓑ	ⓒ	ⓓ
48.	ⓐ	ⓑ	ⓒ	ⓓ
49.	ⓐ	ⓑ	ⓒ	ⓓ
50.	ⓐ	ⓑ	ⓒ	ⓓ

1. Simplify the expression: $6 - {}^{-}9$.

a. -3

b. 15

c. -15

d. 3

2. Simplify the expression: $-3 \cdot -4 \cdot 2$.

a. -24

b. 9

c. -9

d. 24

3. Simplify the expression: $12 - 5 + 3$.

a. -10

b. 4

c. 10

d. -4

4. Simplify the expression: $\frac{50}{-25}$.

a. -2

b. -25

c. 25

d. 2

5. Simplify the expression: $3 + {}^{-}2 \cdot 4 - 4 \div 2$

a. -3

b. 0

c. 2

d. -7

6. Simplify the expression: $3 \cdot 2^2 - 4(2 - 5)$.

a. 0

b. 48

c. 24

d. -1

7. Evaluate $3a + b - 2c^2$ when $a = 2$, $b = -5$, and $c = 4$.

a. 65

b. -63

c. -31

d. 33

8. Simplify: $3x^2y - 10xy + x^2y + 3xy$.

a. $-3x^2y$

b. $-3x^3y^2$

c. $7x^2y + 4xy$

d. $4x^2y - 7xy$

9. Simplify: $5(a + b) - 8(a - b)$

a. $-3a$

b. $-3a - 3b$

c. $-3a + 13b$

d. $-3a + 3b$

10. Solve the equation: $x - 8 = -12$.

a. -4

b. 4

c. -20

d. 20

11. Solve the equation: $x - {}^{-}5 = 7$

a. -2

b. 2

c. 12

d. -12

12. Solve the equation: $-4x = -20$.

a. 5

b. -5

c. 80

d. -80

13. Solve the equation: $\frac{4}{5}x = 20$.

a. 400

b. 25

c. 16

d. 1

14. Solve the equation: $-6x - 3 = 15$.

a. 2

b. 3

c. -3

d. -2

15. Solve the equation: $2x + 5 = -7$.

a. 1
b. –1
c. –6
d. 6

16. Solve the equation: $\frac{1}{5}x + 5 = 25$

a. 100
b. –100
c. 4
d. –4

17. Solve the equation: $4d = 6d - 10$.

a. 1
b. –5
c. –1
d. 5

18. Solve the equation: $6c - 4 = 2c + 16$.

a. $\frac{3}{2}$
b. 3
c. 5
d. –3

19. Convert 50 degrees Fahrenheit into degrees Celsius using the formula $F = \frac{9}{5}C + 32$.

a. 45° C
b. 2° C
c. 10° C
d. 122° C

20. What amount of money would you have to invest to earn \$3,000 in 10 years if the interest rate is 5%? Use the formula $I = prt$.

a. \$1,500
b. \$60,000
c. \$600
d. \$6,000

21. What is the slope in the equation $y = -3x + 5$?

a. $\frac{1}{3}$
b. –3
c. 5
d. 3

22. Transform the equation $6x + 3y = 15$ into slope-intercept form.

a. $3y = 6x + 15$
b. $3y = -6x + 15$
c. $y = -2x + 5$
d. $y = 2x + 5$

23. Choose the equation that fits the graph.

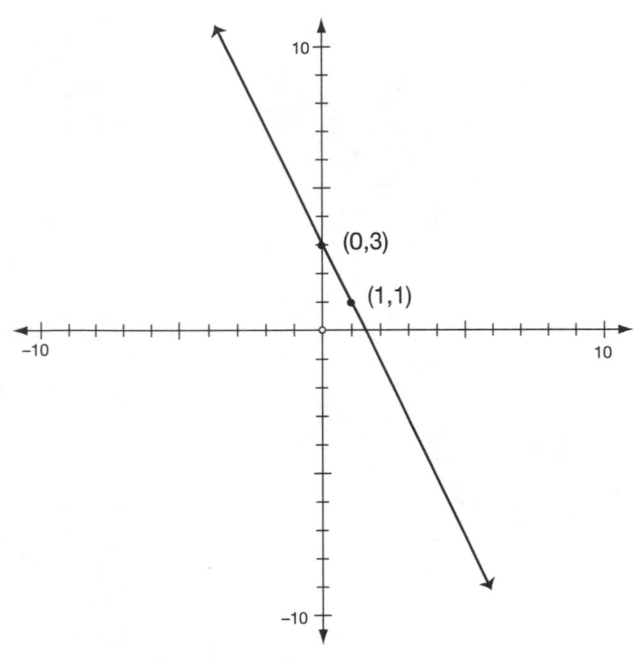

a. $y = 2x + 3$
b. $3x + y = 2$
c. $y = -2x + 3$
d. $-3x + y = 2$

24. Solve the inequality: $5x - 5 < 20$.

a. $x < 5$
b. $x > 5$
c. $x > 3$
d. $x < 3$

25. Solve the inequality: $-4x - 8 > 12$.

a. $x < -5$
b. $x < 5$
c. $x > -1$
d. $x > -5$

26. Match the graph with the inequality: $x > 4$.

a.

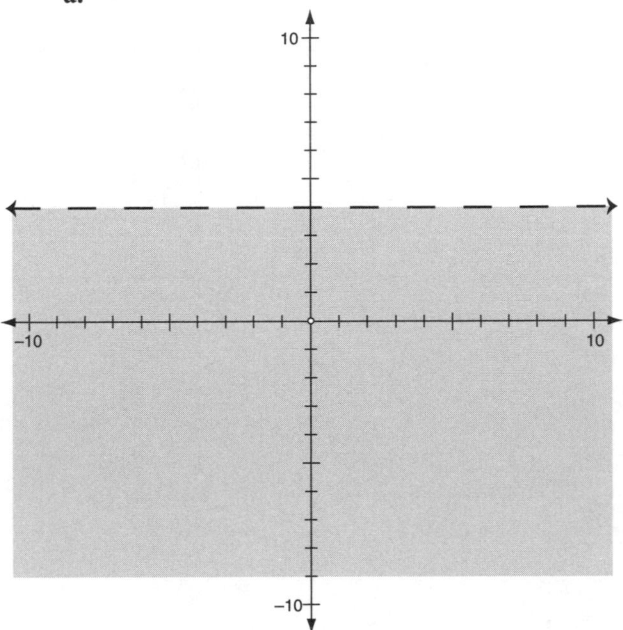

b.

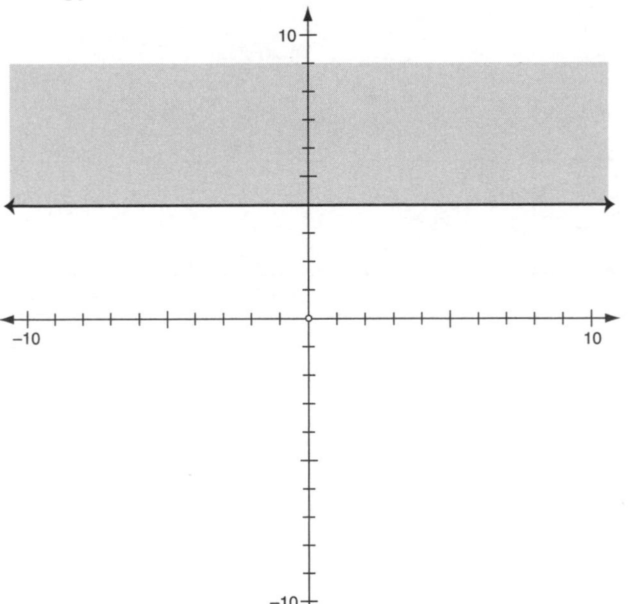

c.

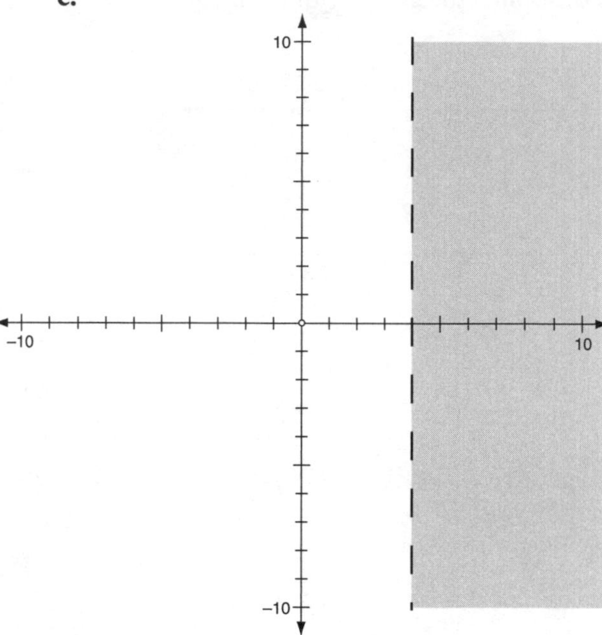

d.

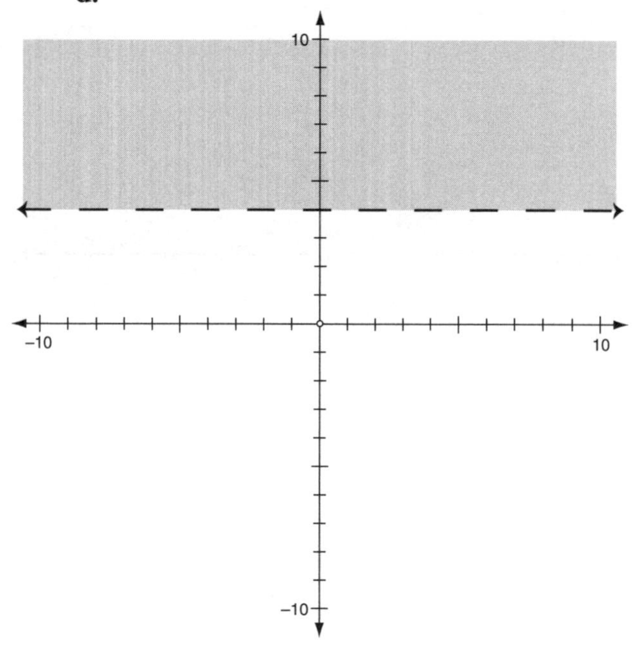

27. Match the inequality with the graph.

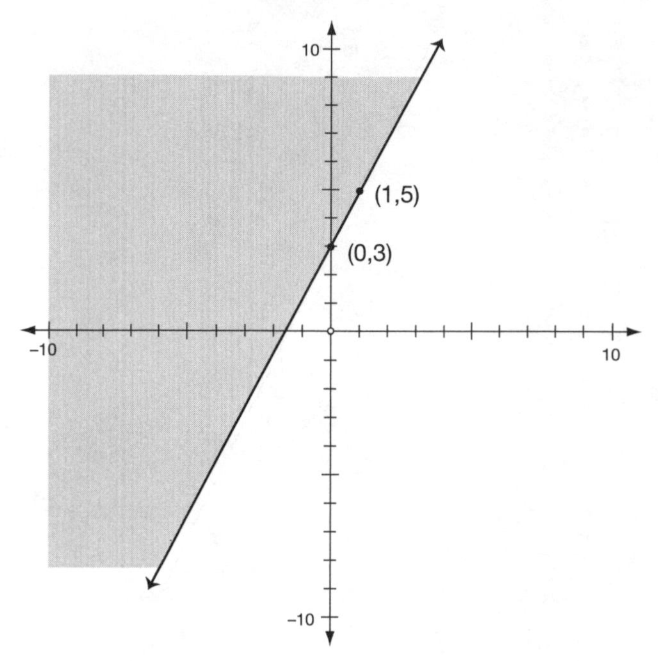

a. $y \leq 2x + 3$
b. $y < 2x + 3$
c. $y > 2x + 3$
d. $y \geq 2x + 3$

28. Determine the number of solutions the system of equations has by looking at the graph.

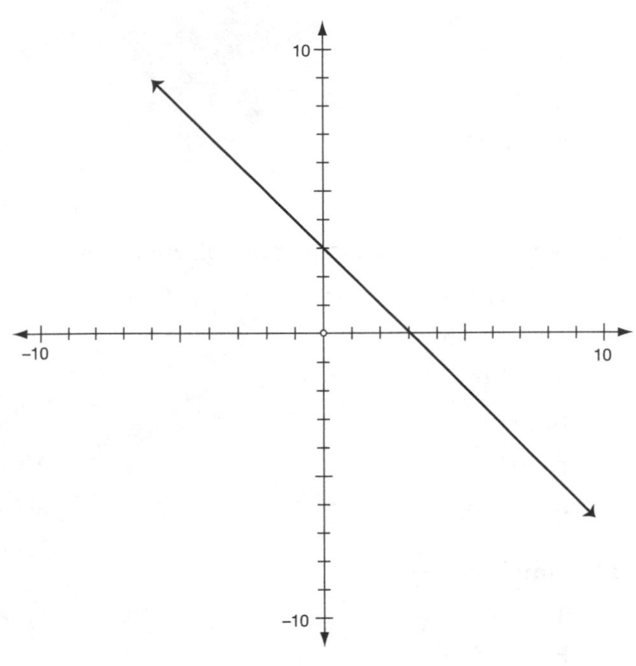

a. 1
b. 0
c. infinite
d. none of the above

29. Determine the number of solutions for the system of equations using the slope and y-intercept.

$y = 3x + 2$

$3x + y = 2$

a. 1
b. 0
c. infinite
d. none of the above

30. Select the graph for the system of inequalities:

$y \geq 2$

$y \leq 2x + 1$

a.

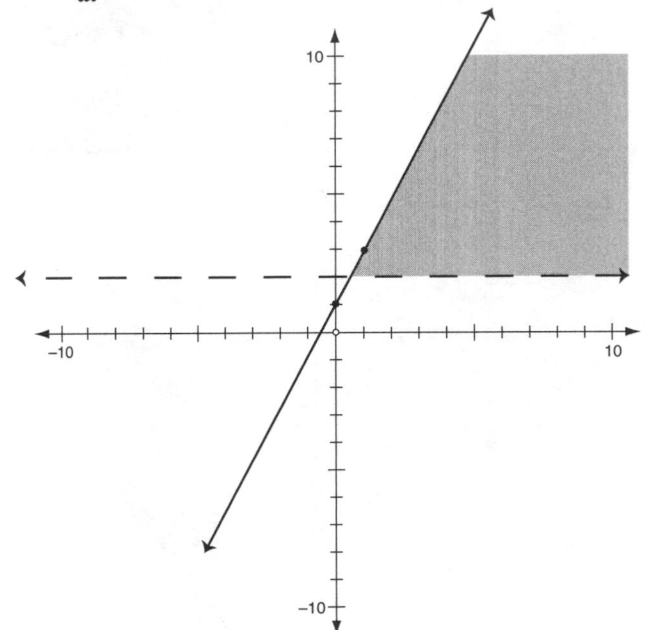

b.

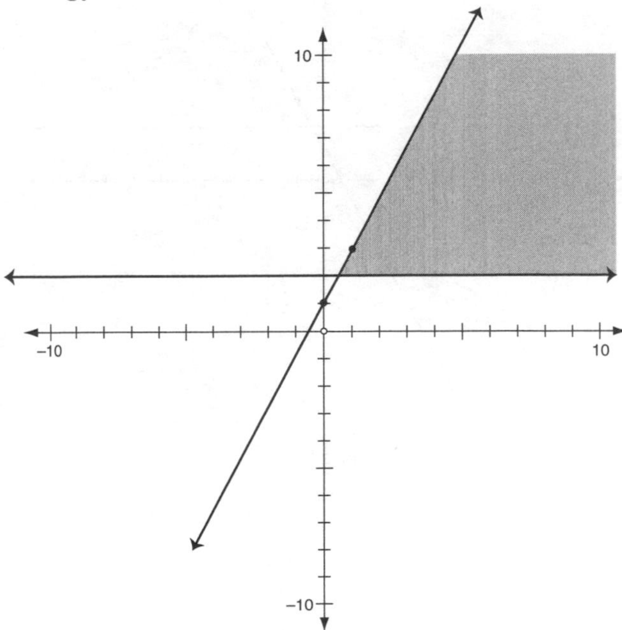

c.

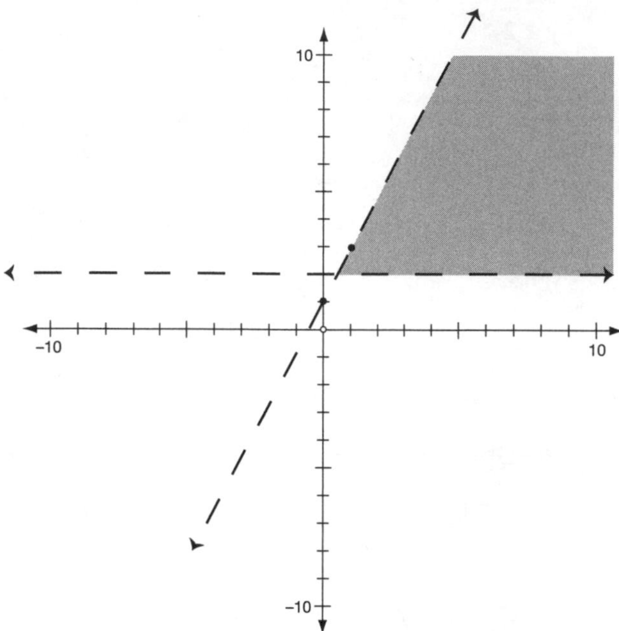

d.

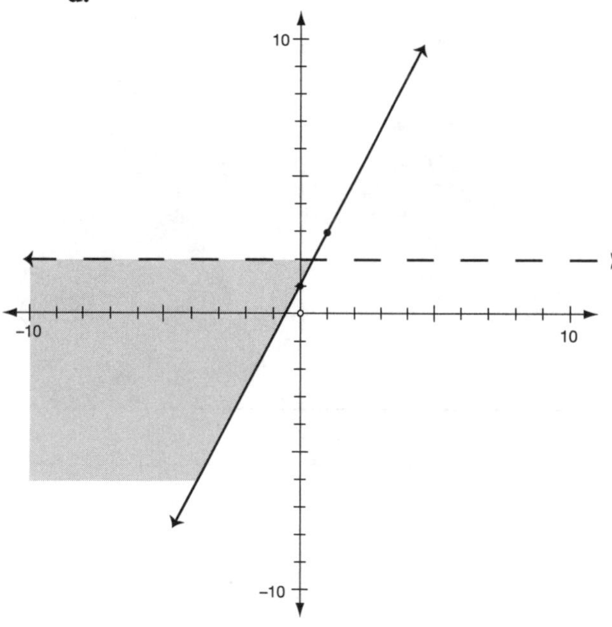

31. Solve the system of equations algebraically:

$-x + 3y = 11$

$x - 5y = -3$

a. (17,4)
b. (1,4)
c. (1,–4)
d. (–23,–4)

32. Solve the system of equations algebraically:

$5x + 2y = 10$

$2x - y = -5$

a. (–5,0)
b. (5,0)
c. (0,5)
d. (0,–5)

33. Simplify: $3x^2y(2x^3y^2)$.

a. $6x^6y^2$
b. $6x^5y^2$
c. $6x^5y^3$
d. $6x^6y^3$

34. Simplify: $4(3x^3)^2$.

a. $12x^5$
b. $144x^6$
c. $12x^6$
d. $36x^6$

35. Multiply the polynomials: $8xy(2x - 3y + 10x^2y)$.

a. $72x^2y^2$
b. $16x^2y - 24xy^2 + 80x^3y^2$
c. $26x^2y - 3y$
d. $96x^2y - 24xy^2$

36. Multiply the binomials: $(5x - 2)(4x + 7)$.

a. $20x^2 - 14$
b. $20x^2 + 20x - 14$
c. $20x^2 + 43x - 14$
d. $20x^2 + 27x - 14$

37. Factor the polynomial: $3ab^2 + 6a^3b^2 - 15a^2b^4$.

a. $3ab^2(2a^2 - 5ab^2)$
b. $3ab(b + 2a^2b - 5ab^3)$
c. $3ab^2(1 + 2a^2 - 5ab^2)$
d. $3ab^2(1 + 2ab - 5ab^2)$

38. Factor the polynomial: $64w^2 - 49$.

a. $(8w - 7)(8w - 7)$
b. $(8w + 7)(8w + 7)$
c. $(8w - 7)(8w + 7)$
d. $(8w - 7)^2$

39. Factor the polynomial: $x^2 + x - 12$.

a. $(x + 2)(x - 6)$
b. $(x - 4)(x + 3)$
c. $(x - 4)(x - 3)$
d. $(x + 4)(x - 3)$

40. Factor the polynomial: $2x^2 + 9x - 5$.

a. $(2x - 1)(x + 5)$
b. $(2x + 5)(x - 1)$
c. $(2x + 1)(x - 5)$
d. $(2x + 5)(x + 1)$

41. Solve the equation: $3x^2 - 75 = 0$.

a. 0,5
b. 5,5
c. 5,–5
d. –5,–5

42. Solve the equation: $2x^2 - x = 3$.

a. $-1,\frac{3}{2}$
b. $-\frac{3}{2},1$
c. $3,-\frac{1}{2}$
d. $-3,\frac{1}{2}$

43. Simplify: $3\sqrt{x} - 2\sqrt{x} + 3\sqrt{y}$.

a. $5\sqrt{xy}$
b. $\sqrt{x} - 3\sqrt{y}$
c. $\sqrt{x} + 3\sqrt{y}$
d. $5\sqrt{x} + 3\sqrt{y}$

44. Simplify: $3\sqrt{2} \cdot 6\sqrt{x}$.

a. $36x$
b. $36\sqrt{x}$
c. $18\sqrt{2x}$
d. $9\sqrt{2x}$

45. Simplify: $2\sqrt{20} - 5\sqrt{45}$.

a. -15
b. $-11\sqrt{5}$
c. $19\sqrt{5}$
d. $-3\sqrt{-25}$

46. Simplify: $\sqrt{\frac{5x}{3}}$.

a. $\frac{5}{3}\sqrt{x}$
b. $\frac{1}{3}\sqrt{3x}$
c. $\frac{1}{3}\sqrt{15x}$
d. $\frac{1}{3}\sqrt{5x}$

47. Solve the equation: $2\sqrt{x} - 1 = 15$.

a. 7
b. 8
c. 49
d. 64

48. Solve the equation: $2\sqrt{x - 3} + 4 = 8$.

a. 4
b. 7
c. 39
d. 67

49. Use the quadratic formula to solve:
$3x^2 - 7x + 2 = 0$.
a. $\frac{1}{3}$, 2
b. $\frac{1}{3}$, -2
c. $-\frac{1}{3}$, -2
d. $-\frac{1}{3}$, -2

50. Use the quadratic formula to solve:
$4x^2 + 7x + 1 = 0$.
a. $\frac{7 \pm \sqrt{65}}{8}$
b. $\frac{-7 \pm \sqrt{65}}{8}$
c. $\frac{7 \pm \sqrt{33}}{8}$
d. $\frac{-7 \pm \sqrt{33}}{8}$

Answer Key

Pretest

If you miss any of the answers, you can find help for that kind of question in the lesson shown to the right of the answer.

1. b (1)
2. d (1)
3. b (1)
4. d (1)
5. a (2)
6. b (2)
7. a (2)
8. c (3)
9. c (3)
10. a (4)
11. a (4)
12. b (4)
13. c (4)
14. b (5)
15. d (5)
16. b (5)
17. c (6)
18. b (6)
19. a (7)
20. d (7)
21. b (8)
22. b (8)
23. a (8)
24. b (9)
25. c (9)
26. d (10)
27. c (10)
28. b (11)
29. a (11)
30. a (11)
31. b (12)
32. b (12)
33. c (13)
34. d (13)
35. b (14)
36. c (14)
37. d (15)
38. a (15)

39. d (16)
40. a (16)
41. c (17)
42. b (17)
43. c (18)
44. c (18)
45. c (18)
46. c (18)
47. c (19)
48. b (19)
49. b (20)
50. b (20)

Lesson 1

1. <
2. >
3. <
4. >
5. <
6. >
7. >
8. >
9. 12
10. –12
11. –8
12. –11
13. 7
14. –6
15. 9
16. 0
17. –1
18. 9
19. –7
20. –19
21. –12
22. –13
23. –6
24. –5
25. 7
26. 12
27. 0
28. –18
29. 5
30. –6
31. 12
32. 14
33. –2
34. –11
35. 1
36. 0
37. 8
38. 4
39. 10
40. 8
41. 3
42. 19
43. –17
44. 5
45. 4
46. 8
47. 13
48. –3
49. 56
50. –20
51. –7
52. 24
53. 7
54. –11
55. 4
56. –8
57. –77
58. 25
59. –24
60. 40
61. –36
62. 90

63. –48
64. 80
65. –5
66. –10
67. –3
68. 3
69. 21
70. 2
71. –56
72. –47
73. 27
74. –26
75. –13
76. –15
77. –60
78. 0
79. 25
80. –12
81. –42
82. 3
83. 60
84. 16
85. 20
86. 16
87. –6
88. –12
89. 400
90. –25
91. 85° – 17° = 68°
92. 30° – $^-$7° = 37°
93. \$45 – \$55 = –\$10
94. –\$20 + \$100 = \$80
95. –4 in., a loss of 4 inches

Lesson 2

1. 16
2. 6
3. –7
4. 21
5. 11
6. 9
7. 10
8. 16
9. 3
10. 2
11. 15
12. 17
13. –3
14. –4
15. 40
16. 7
17. –17
18. –43
19. –2
20. –4
21. 10
22. 8
23. 14
24. –33
25. 10
26. –37
27. 16
28. –32
29. 36
30. 18
31. 12
32. 3
33. 28
34. 5
35. 13
36. 80
37. 25

38. –33
39. 0
40. 16

Lesson 3

1. $17x$
2. $13x + 7y$
3. $15x + 5$
4. $18x$
5. $2x^2 - 5x + 12$
6. $5x - 7y + 2$
7. $12x + 12y + 5$
8. $8x^2 + 5x + 9$
9. 8
10. $7xy + 16x - 2y - 8$ (Note: $xy = yx$)
11. $6x + 2y$
12. $8a + 8b$
13. $8m + 31$
14. $3x - 8y$
15. $12r + 3s$
16. $11m + n - 2$
17. $13x + 12y$
18. $5x - 16$
19. $-13a - 7b$
20. $17x + 13y$
21. –8
22. 2
23. 6
24. –22
25. 0
26. –1
27. –18
28. 0
29. –30
30. 0
31. –9
32. 19
33. –5
34. 0
35. 25
36. 80
37. 6
38. 5

Lesson 4

1. –6
2. 15
3. –1
4. –7
5. 5
6. 5
7. 4
8. 8
9. 17
10. –17
11. 12
12. 48
13. 9
14. 128
15. 26
16. $\frac{4}{3} = 1\frac{1}{3}$
17. $x = \frac{5}{3}$ or $x = 1\frac{2}{3}$
18. 2
19. $\frac{4}{3} = 1\frac{1}{3}$
20. $\frac{1}{3}$
21. 8
22. $4\frac{1}{2}$
23. $\frac{10}{3} = 3\frac{1}{3}$
24. 21
25. 3
26. –15
27. –18
28. 42
29. –14
30. –75
31. –8

32. –63
33. –13
34. –6
35. 30
36. –29
37. $\frac{1}{10}$
38. $-\frac{3}{4}$
39. –7
40. –40
41. $x = \frac{\$1{,}000 - \$525}{24 \text{ hours}}$, \$19.79 per hour
42. $x = 0.15 \cdot \$32$, $x = \$4.80$
43. $40 \cdot \frac{x}{35} = \$560$, $x = \$490$
44. $x = .90 \cdot 40$, 36 questions
45. $.80x = 32$, 40 questions
46. $\$7.50x = \$30$, 4 hours
47. $x + x + 2x + 2x = 30$, width = 5, length = 10
48. $x = .25 \cdot \$22{,}000$, \$5,500 per year, \$458.33 per month
49. $4x \cdot 0.05 \cdot 4x = \$46.20$, $x = \$11$
50. $400 = 6r$, 66.7 miles per hour

Lesson 5

1. 1
2. 5
3. 2
4. –5
5. 3
6. 63
7. –7
8. $-\frac{5}{2}$
9. –2
10. –5
11. 5
12. –6
13. $\frac{5}{3} = 1\frac{2}{3}$
14. –40
15. 1
16. –52
17. –13
18. 1
19. –4
20. 0
21. $\frac{2}{3}x = \$38{,}000$, \$57,000
22. $\frac{4}{5}x = \$400$, \$500
23. $2x + 5 = 35$, 15 sit-ups
24. $x + (2x + 3) = 27$, $x = 8$ boys
25. $x + 0.05x + 500 = 1{,}760$, $x = \$1{,}200$
26. 21
27. 12
28. 4
29. 10
30. –24
31. 2
32. 0
33. –15
34. $\frac{25}{6} = 4\frac{1}{6}$
35. –4.5

Lesson 6

1. –2
2. 5
3. –2
4. $x = 8$
5. –3
6. 2
7. $-\frac{2}{3}$
8. –6
9. 2
10. 6
11. 2
12. –9
13. 7
14. 4
15. 4
16. 0
17. $-\frac{9}{4} = -2\frac{1}{4}$

18. -1

19. 0

20. $-\frac{6}{8} = -\frac{3}{4}$

21. -1

22. 4

23. -2

24. 1

25. -2

26. 3

27. -3

28. -5

29. 0

30. $-\frac{1}{3}$

31. No solution or $\varnothing$

32. Any real number or R

33. Any real number or R

34. -3

35. No solution or $\varnothing$

36. $\frac{1}{2}x + \$200 = \$450$, \$500

37. $\$15 + 1\%(\$2{,}365) = x$, \$38.65

38. $x + x + (3 + 2x) + (3 + 2x) = 66$, width = 10, length = 23

39. $45 + 10x = 180$, $x = 13.15$ lb.

40. $8 \cdot 2x + \$50 = \$218$, \$10.50

Lesson 7

1. $A + 24.2$ cm $\cdot$ 14 cm, 338.8 cm^2

2. 360 ft.$^2 = l \cdot 12$ ft., 30 ft.

3. $I = \$2{,}000 \cdot 11.5\% \cdot 2$ yrs., \$460

4. $\$138 = p \cdot 11.5\% \cdot 2$ yrs., \$600

5. $\$135 = \$1{,}500 \cdot 3\%t$, 3 years

6. $V = 8 \cdot 5 \cdot 7$, 280 ft.^{3}

7. 270 cm$^3 = l \cdot 6$ cm $\cdot$ 3 cm, 15 cm

8. 858 ft.^{3} = 13 ft. $\cdot$ 11 ft. $\cdot\, h$, 6 ft.

9. $C = 60 \text{ cc} \cdot \frac{50 \text{ lb.}}{150}$, 20 cc

10. $C = 70 \text{ cc} \cdot \frac{90 \text{ lb.}}{150}$, 42 cc

11. $80 \text{ cc} = 80 \text{ cc} \cdot \frac{w}{150}$, 150 lb.

12. $F = 5 \cdot 9.8$, $F = 49$ Newtons

13. $100 = m \cdot 9.8$, $m = 10.2$ kg (approximately)

14. $10{,}000 = 50 \cdot a$, $a = 200\frac{m}{s^2}$

15. $E = 5{,}000 \cdot 12 + 15{,}000$, $E = \$75{,}000$

16. $35{,}000 = 100{,}000 \cdot m + 15{,}000$, $m = \$0.20$ or 20¢

17. $25{,}000 = 500 \cdot 30 + R$, $R = \$10{,}000$

18. $S = -2.817 \cdot 8 + 108.9$, 86.36 decibels

19. $S = -2.817 \cdot 3 + 108.9$, 100.45 decibels

20. $105 = -2.817H + 108.9$, 1.38 hours

21. $100 = \frac{9}{5}C + 32$, 37.78° C

22. $F = \frac{9}{5} \cdot 100 + 32$, 212° F

23. $32 = \frac{9}{5}C + 32$, 0° C

24. $A = \frac{1}{2}(4 + 10) \cdot 5$, A = 35 ft.^{2}

25. $48 = \frac{1}{2}(4 + 12) \cdot h$, $h = 6$ in.

26. $42 = \frac{1}{2}(b_1 + 9) \cdot 7$, $b_1 = 3$ ft.

27. $80 = \frac{1}{2}(x + x) \cdot 10$, $x = 8$ cm

28. $D = \frac{\$29{,}999 - \$600}{20 \text{ years}}$, \$1,469.95

29. $\$29{,}999 - 8 \cdot \$1{,}469.95$, \$18,239.40

30. $D = \frac{\$9{,}500 - \$300}{10 \text{ years}}$, \$920

Lesson 8

1–10.

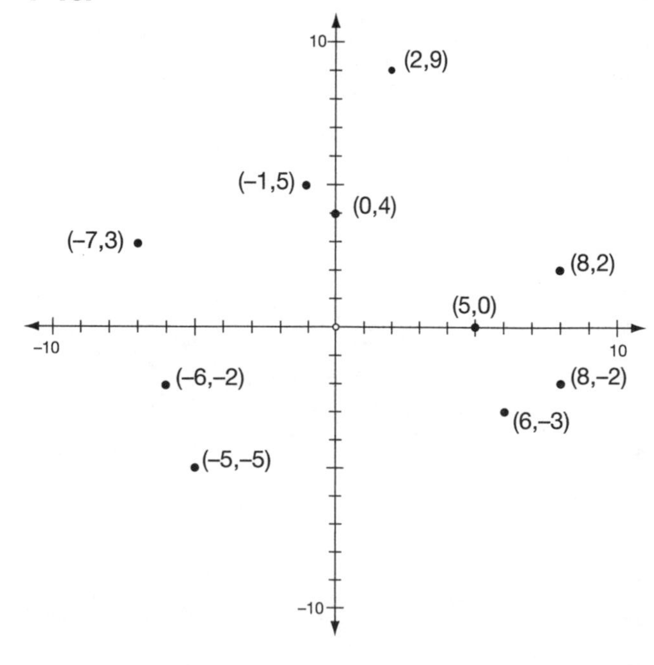

11. $m = 3, b = 9$

12. $m = 5, b = -6$

13. $m = -5, b = 16$

14. $m = -2.3, b = -7.5$

15. $m = \frac{3}{4}, b = 5$

16. $m = \frac{1}{3}, b = 8$

17. $y = -2x - 4, m = -2, b = -4$

18. $y = -3x + 6, m = -3, b = 6$

19. $y = 3x + 8, m = 3, b = 8$

20. $y = -x + 3, m = -1, b = 3$

21. $y = 2x + 5, m = 2, b = 5$

22. $y = -\frac{3}{4}x + 3, m = \frac{3}{4}, b = 3$

23. $y = 4x - 3, m = 4, b = -3$

24. $y = \frac{1}{2}x + 2, m = \frac{1}{2}, b = 2$

25. $y = -2x - \frac{5}{3}, m = -2, b = -\frac{5}{3}$

26. $y = \frac{4}{5}x, m = \frac{4}{5}, b = 0$

27.

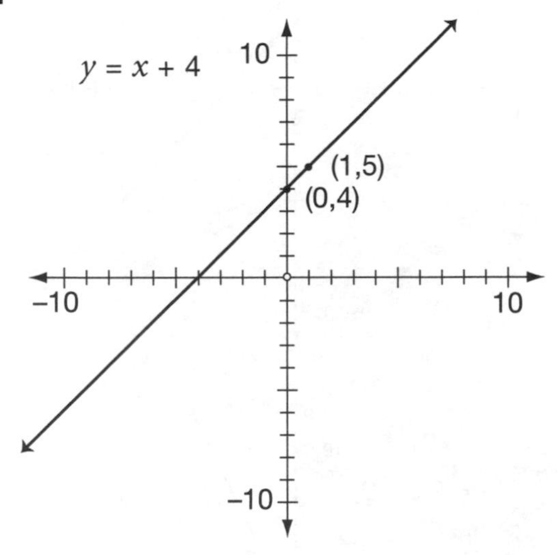

28.

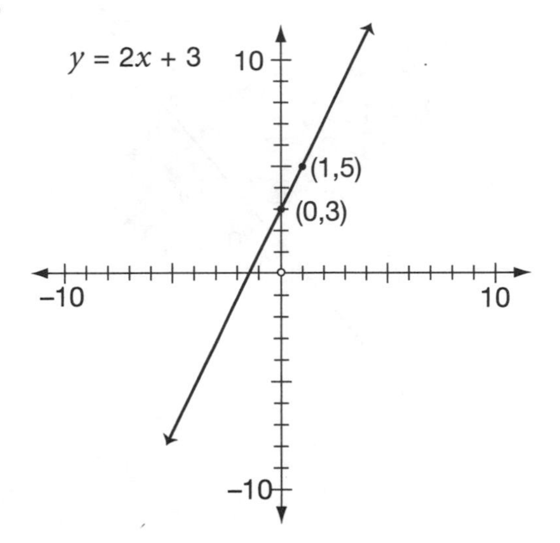

29.

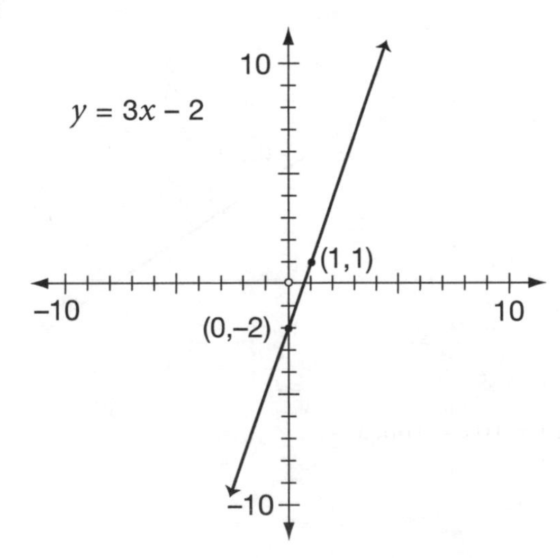

30.

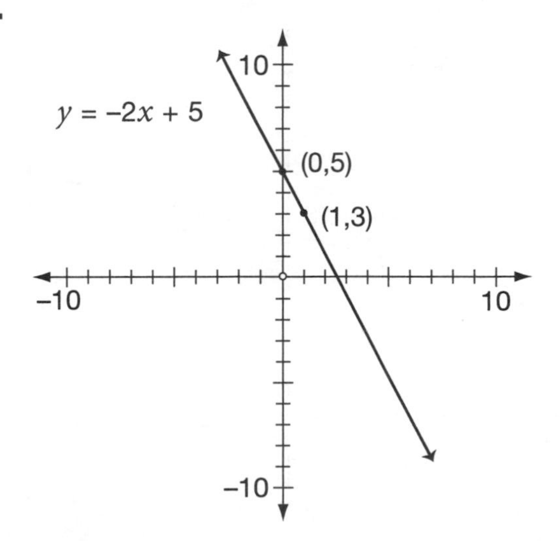

31.

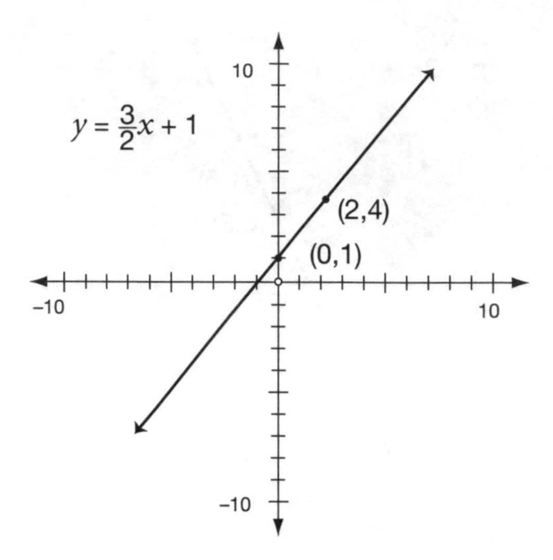

32.

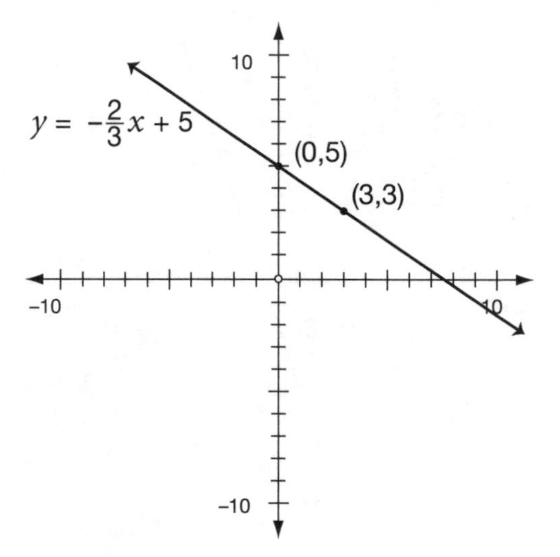

33.

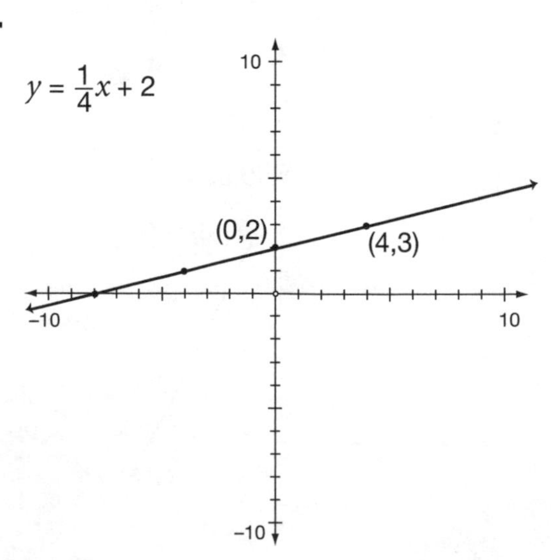

34.

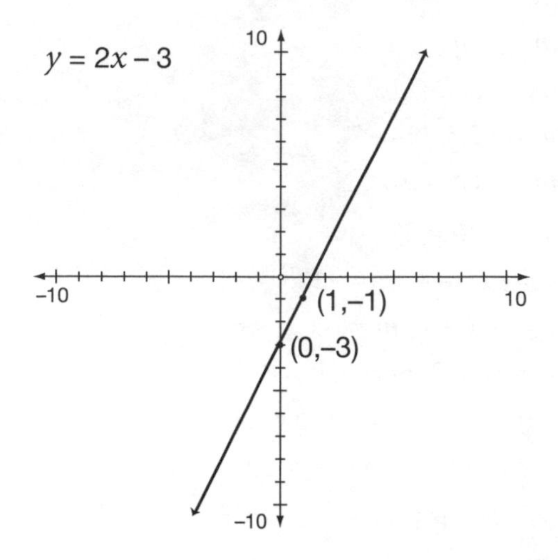

35.

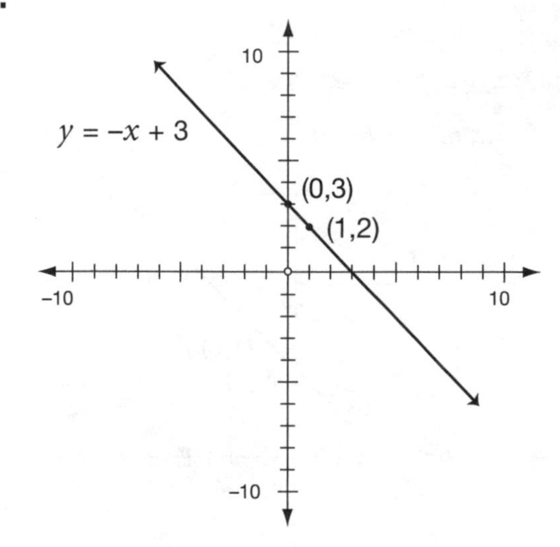

36.

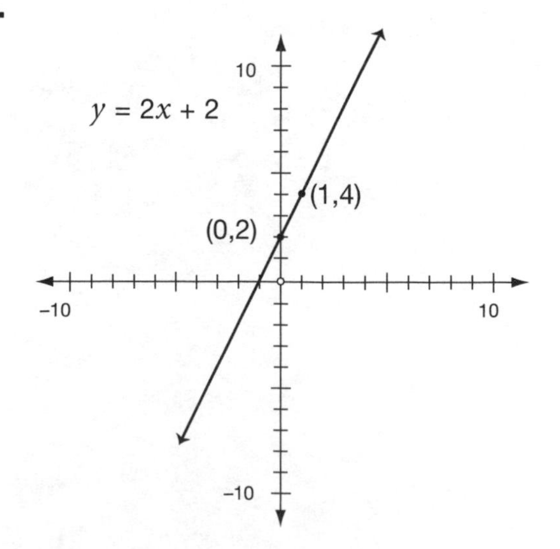

37.

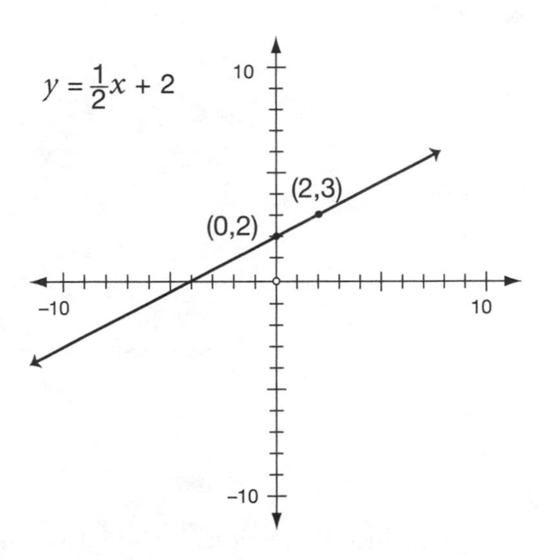
$y = \frac{1}{2}x + 2$
10
(2,3)
(0,2)
–10
10
–10

38.

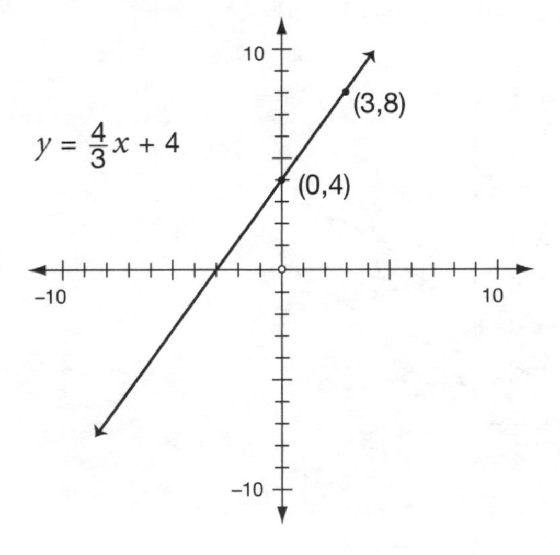
10
$y = \frac{4}{3}x + 4$
(3,8)
(0,4)
–10
10
–10

39.

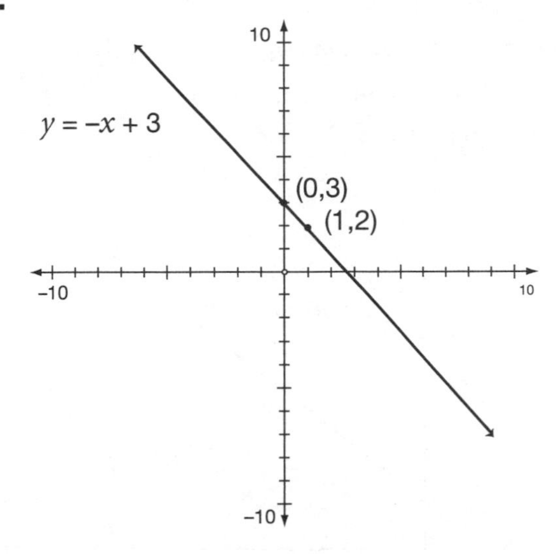
10
$y = -x + 3$
(0,3)
(1,2)
–10
10
–10

40.

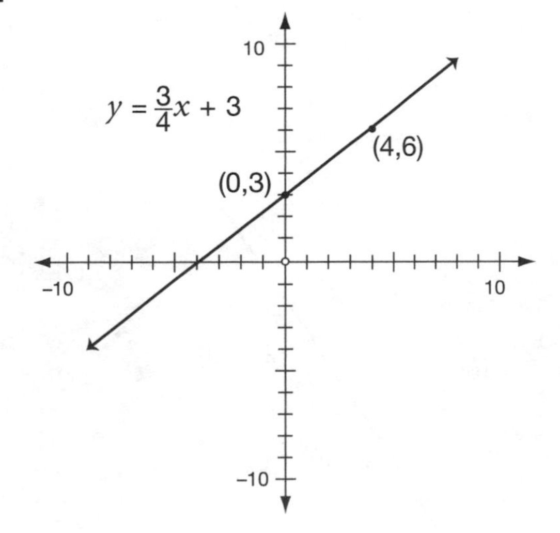
10
$y = \frac{3}{4}x + 3$
(4,6)
(0,3)
–10
10
–10

41. $-\frac{70}{700} = -\frac{1}{10}$

42. $y = 5\% \cdot x + 320$, $m = \frac{5}{100} = \frac{1}{20}$, $b = 320$

43. $y = .10x + 25$, $m = .10$, $b = 25$

44. $y = 0.05 + 5$

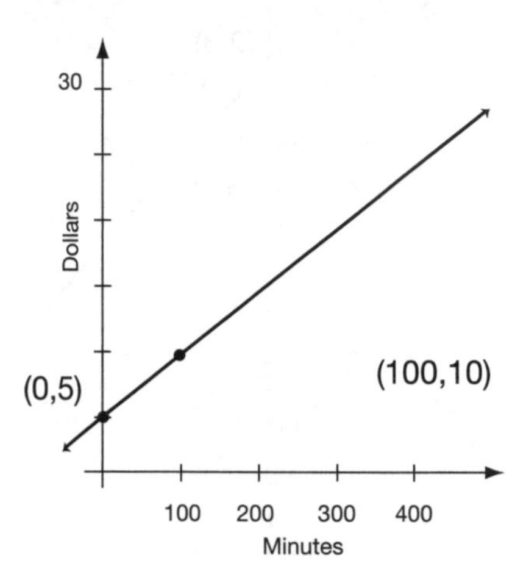

45.

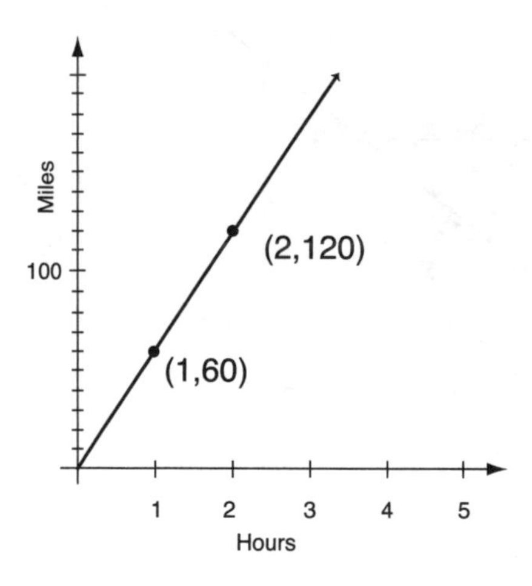

Lesson 9

1. $x < 7$

2. $x \geq 7$

3. $x < 1$

4. $x \leq -\frac{7}{5}$ or $x \leq -1\frac{2}{5}$

5. $x \geq -3$

6. $x \geq -8$

7. $x < 15$

8. $x > -8$

9. $x < 12$

10. $x \leq 2$

11. $x > -6$

12. $x > -\frac{16}{6}$ (Reduce all fractions.) $x > -\frac{8}{3}$ (Another possible answer is $x > -2\frac{2}{3}$.)

13. $x \geq 3$

14. $x \leq 48$

15. $x < -2$

16. $x \leq \frac{3}{2}$ or $x \leq 1\frac{1}{2}$

17. $x > 0$

18. $x \geq \frac{9}{4}$ or $x \geq 2\frac{1}{4}$

19. $x < -8$

20. $x < 0$

21. $x \geq 2$

22. $x \leq -3$

23. $x > 1$

24. $x < 16$

25. $x < -4$

26. $x > -4$

27. $2x + 8 \leq \$20$, $x \leq 6$

28. $4x \leq 120$, $x \leq \$30$; the most you can spend on a chair is \$30.

29. $x + \$2.50 + \$2 \leq \$15$, $x \leq \$10.50$

30. $5x \leq 120$, $x \leq 24$ minutes or $x \leq \frac{2}{5}$ hour

Lesson 10

1.

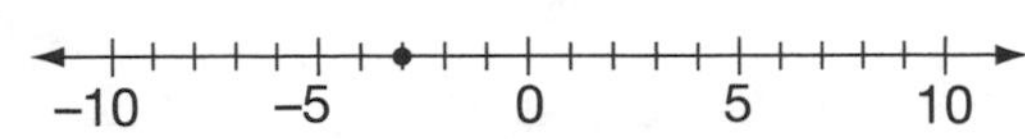

2.

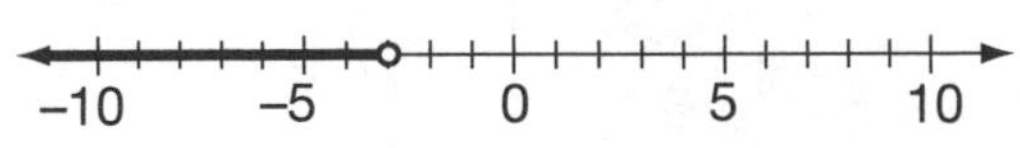

3.

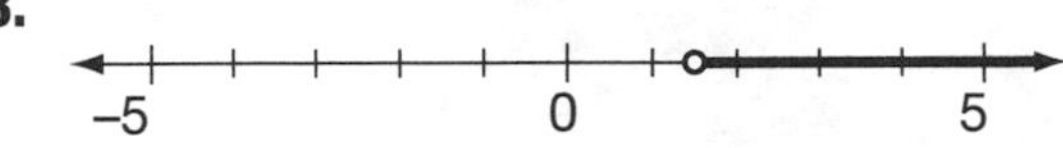

4.

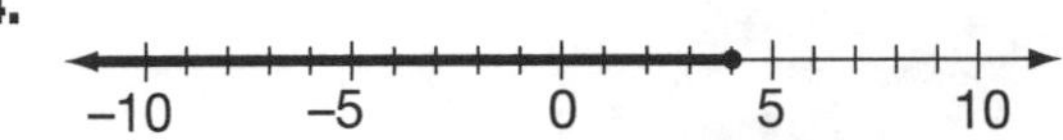

5.

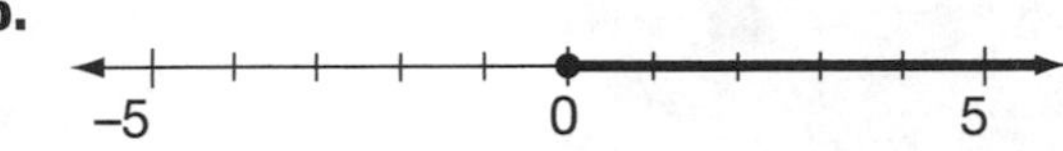

6.

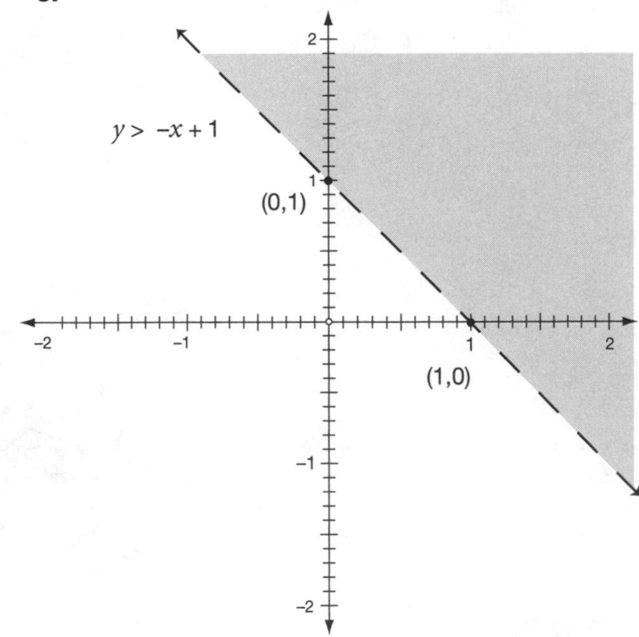

7.

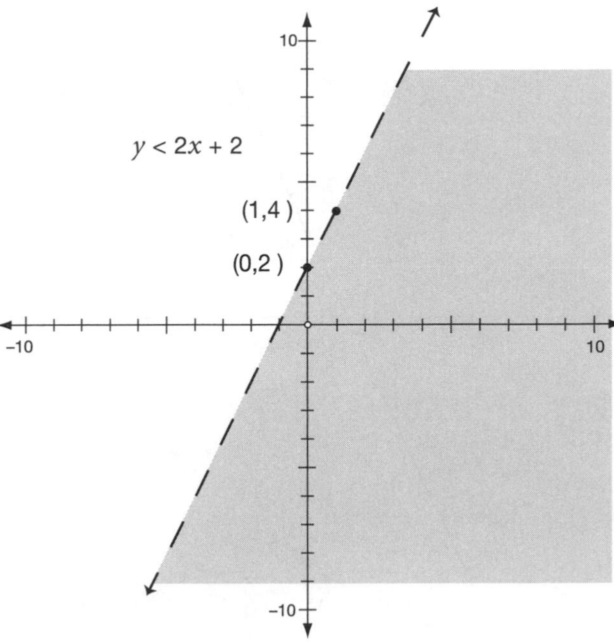

8.

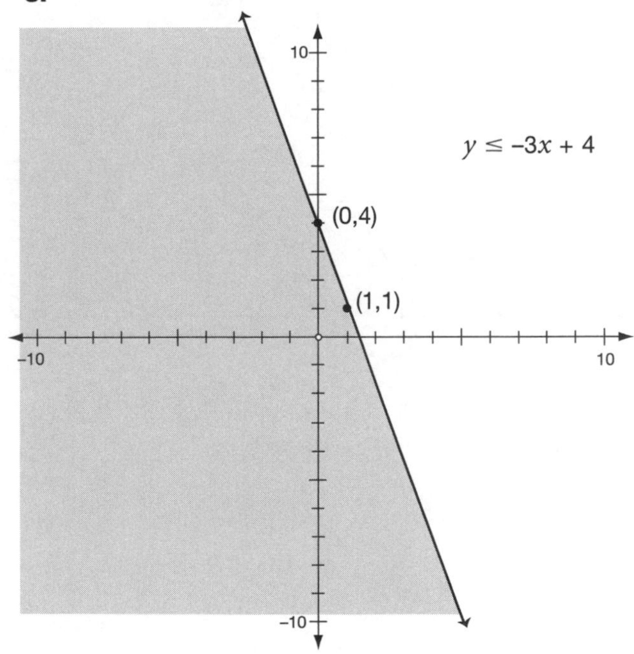

9.

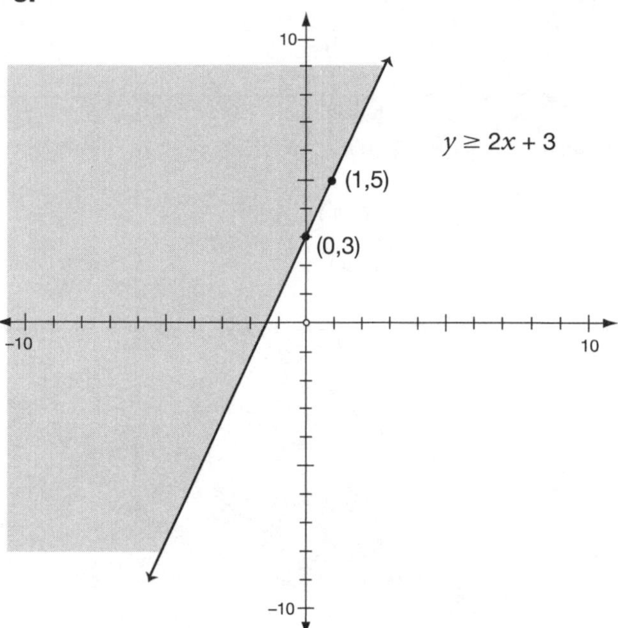
10
-10
-10
10
$y \geq 2x + 3$
(1,5)
(0,3)

10.

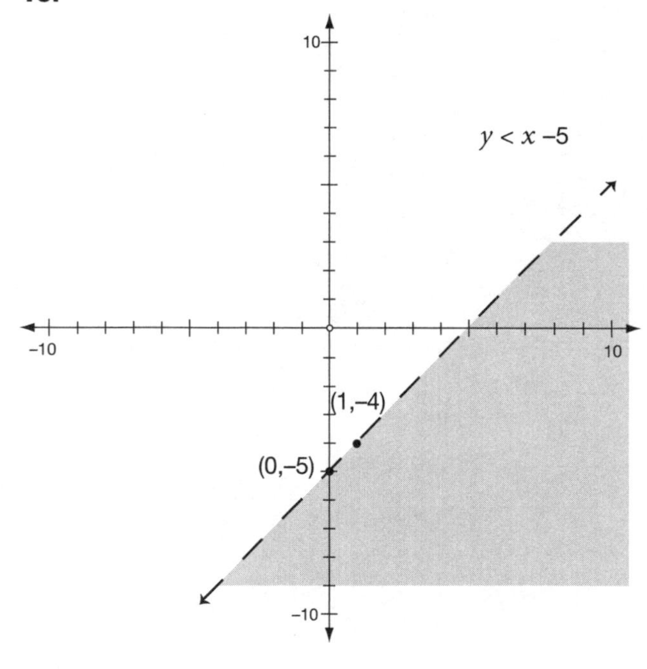
10
-10
-10
10
$y < x - 5$
(1,–4)
(0,–5)

11.

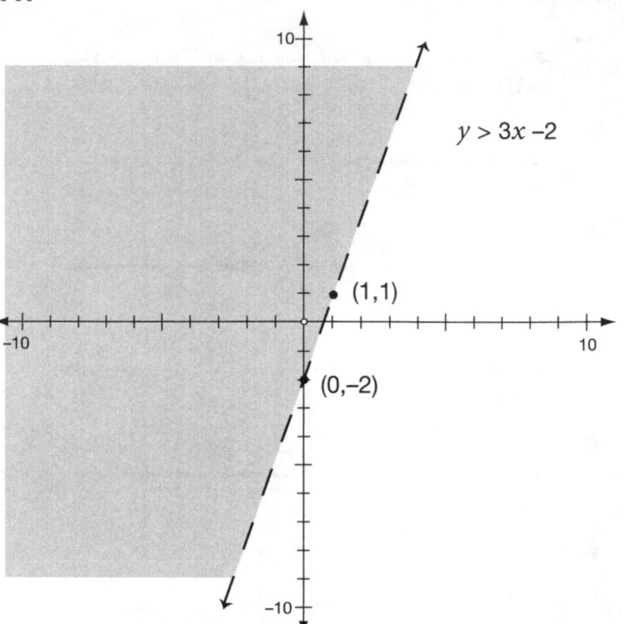
10
-10
-10
10
$y > 3x - 2$
(1,1)
(0,–2)

12.

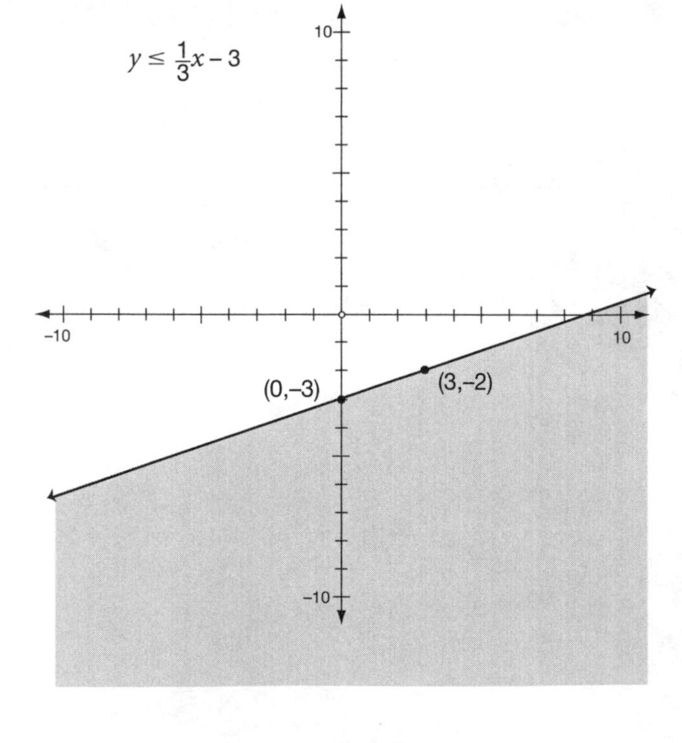
10
-10
-10
10
$y \leq \frac{1}{3}x - 3$
(0,–3)
(3,–2)

13.

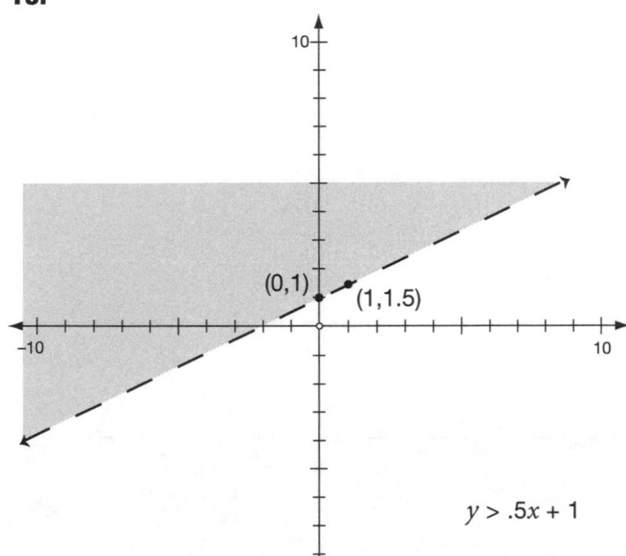
10
−10
−10
10
(0,1)
(1,1.5)
$y > .5x + 1$

14.

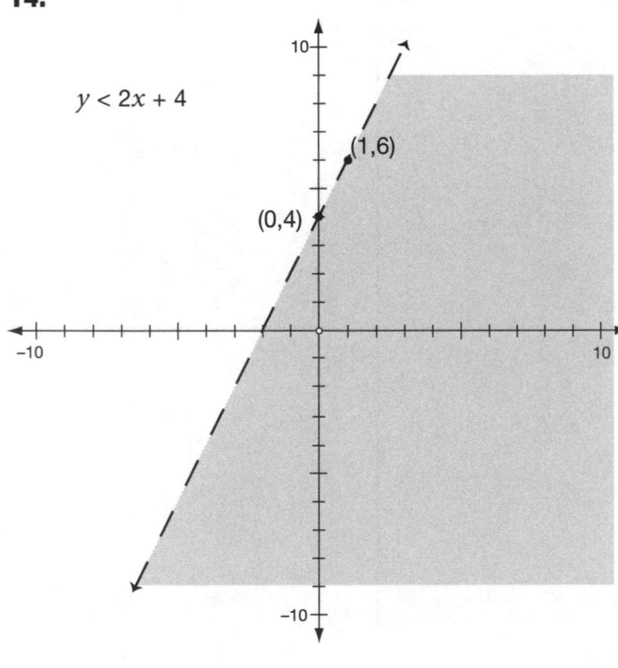
10
−10
−10
10
$y < 2x + 4$
(1,6)
(0,4)

15.

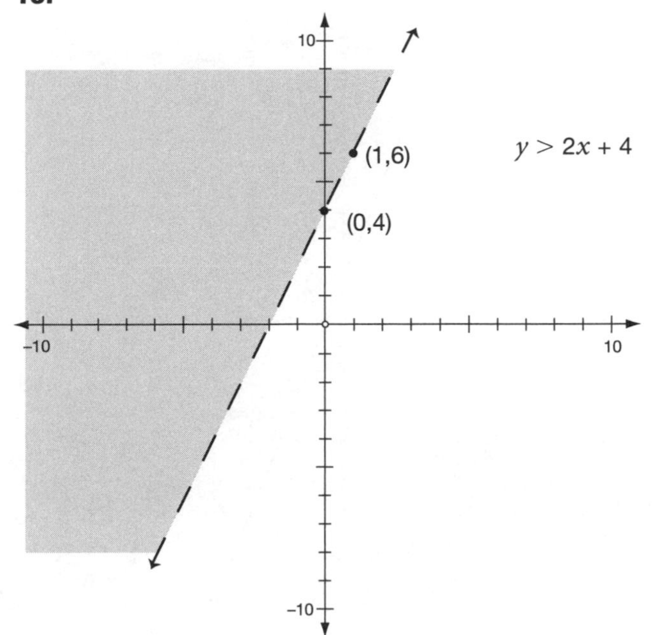
10
−10
−10
10
(1,6)
(0,4)
$y > 2x + 4$

16.

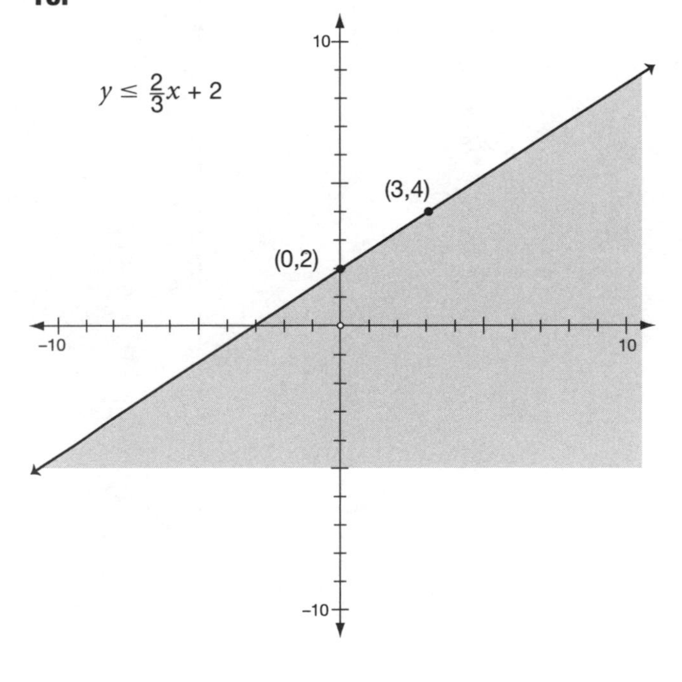
10
−10
−10
10
$y \leq \frac{2}{3}x + 2$
(3,4)
(0,2)

17.

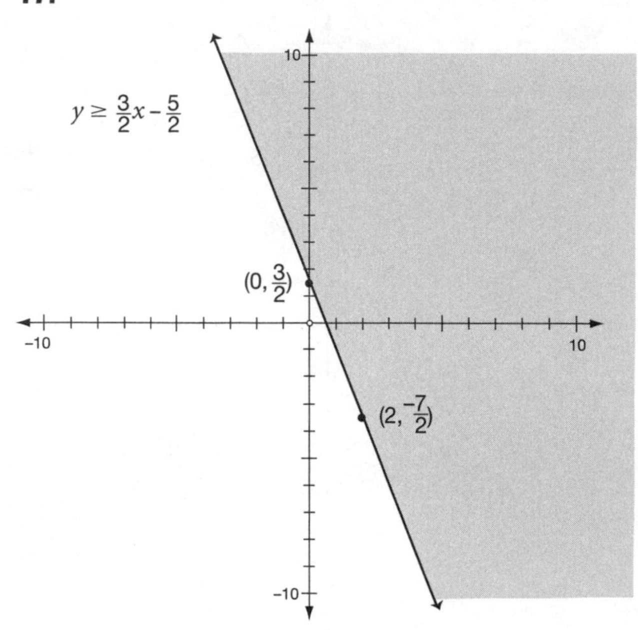

$y \geq \frac{3}{2}x - \frac{5}{2}$
10
−10
−10
10
$(0, \frac{3}{2})$
$(2, \frac{-7}{2})$

19.

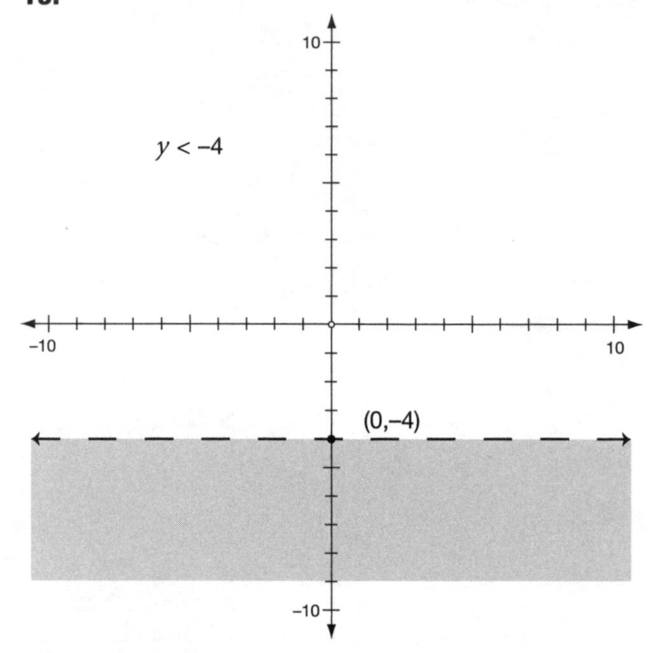

$y < -4$
10
−10
10
−10
(0,−4)

18.

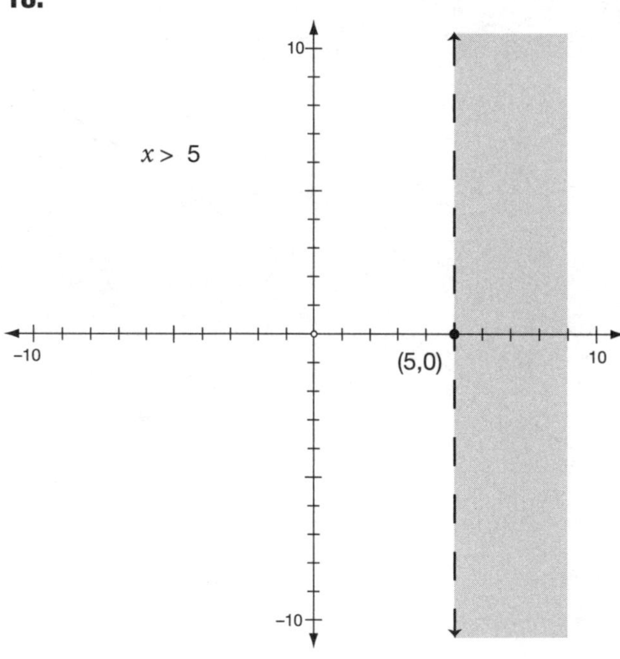

$x > 5$
10
−10
10
−10
(5,0)

20.

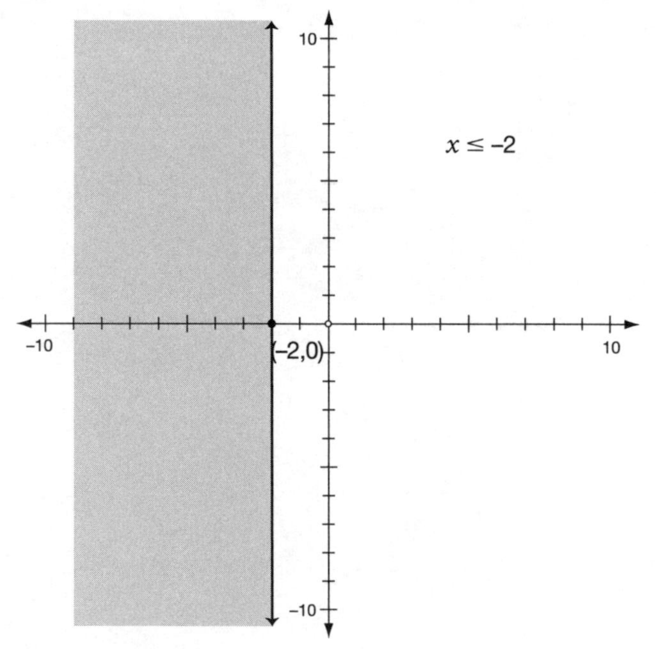

$x \leq -2$
10
−10
10
−10
(−2,0)

21.

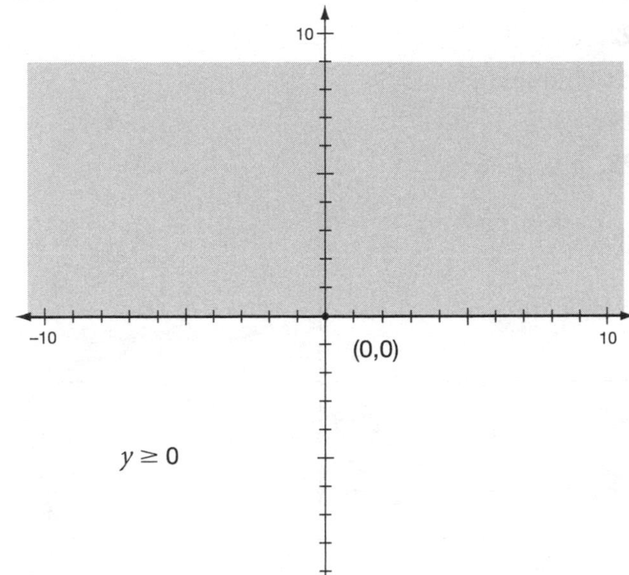

22. $15x + 10y \leq 100$

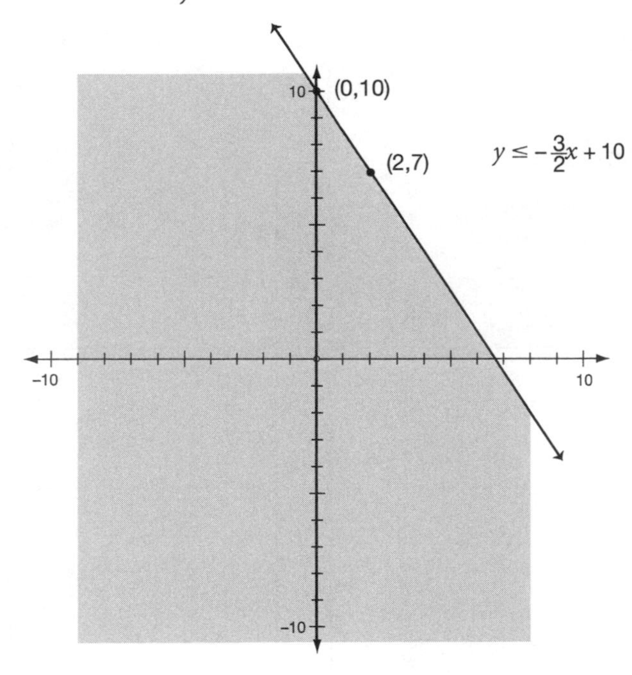

23. $6x + 5y \geq 100$

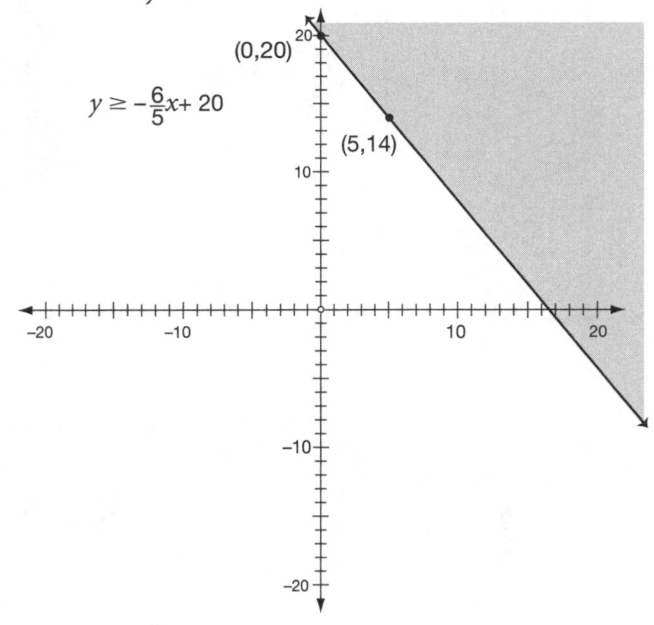

24. $2x + 1\frac{1}{2}y \leq 6$

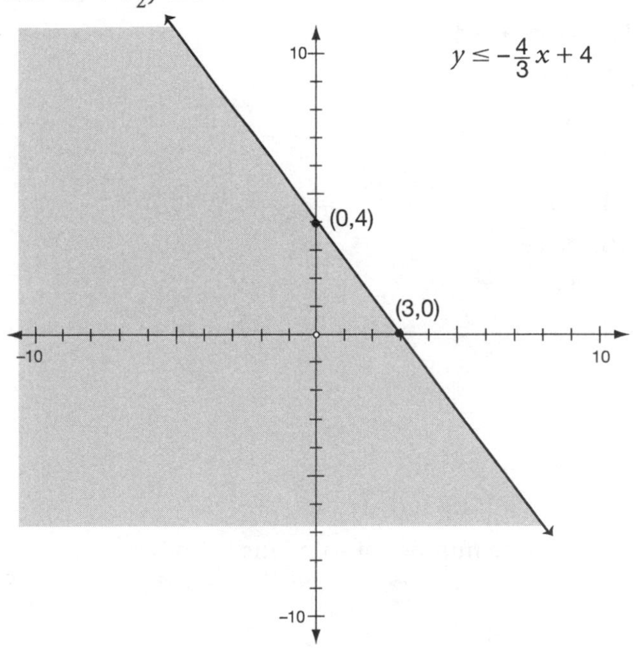

25. $2x + y \geq 20$

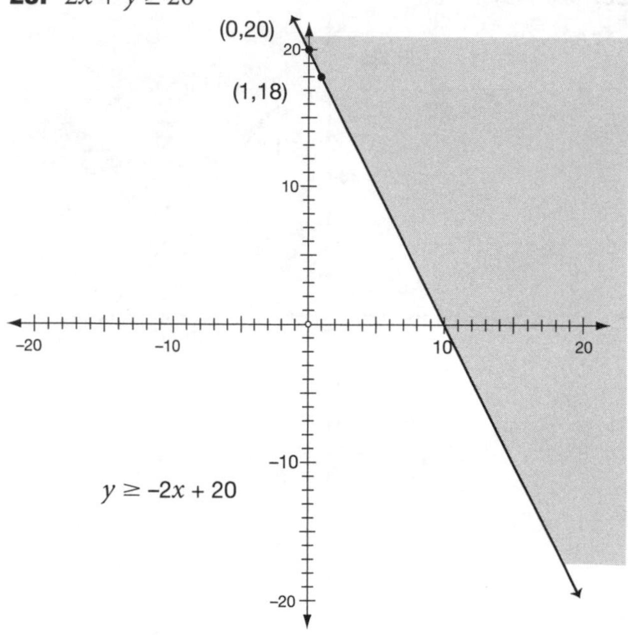

Lesson 11

1. Yes
2. Yes
3. Yes
4. No
5. No
6. Yes
7. No
8. No
9. No
10. Yes
11. One solution, (0,4)
12. Infinite number of solutions
13. No solutions, $\varnothing$
14. One solution, (1,1)
15. No solutions, $\varnothing$
16. One solution, (–3,2)
17. $\varnothing$
18. Infinite
19. One
20. One
21. $\varnothing$
22. One
23. Infinite
24. $\varnothing$
25.

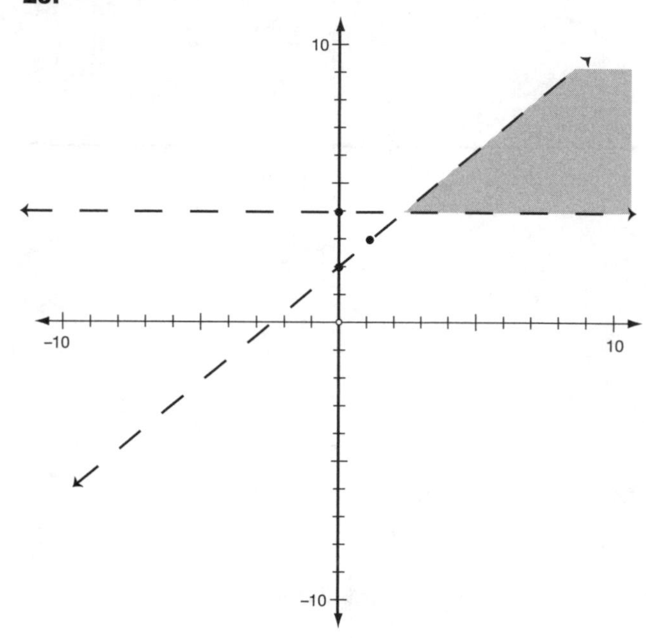

26.

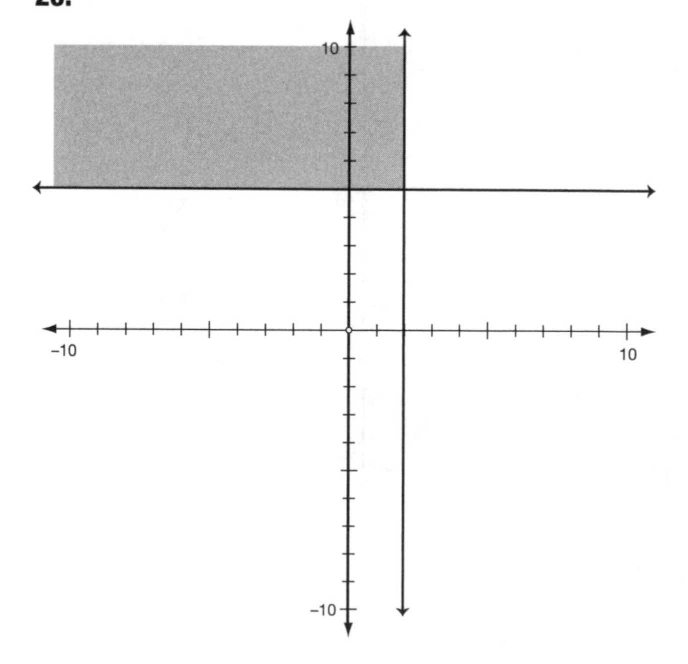

27.

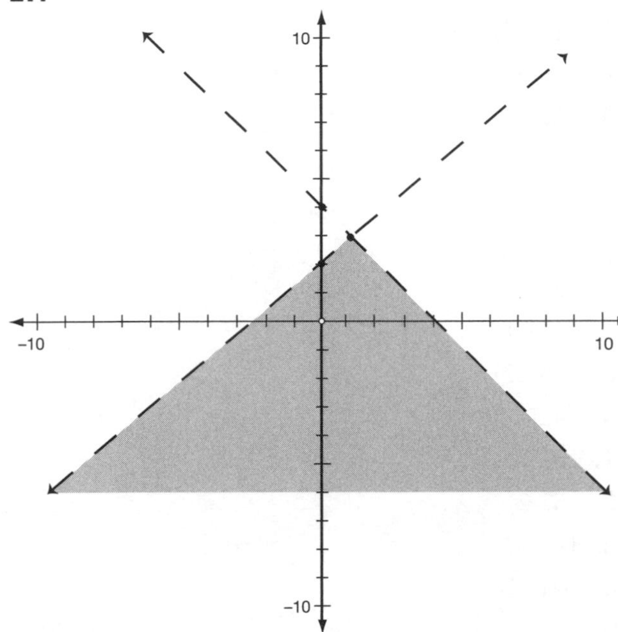

29.

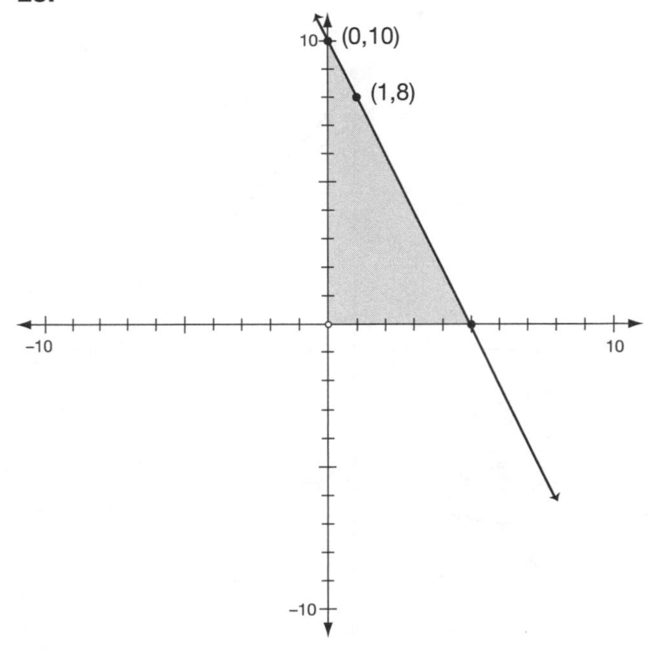

28.

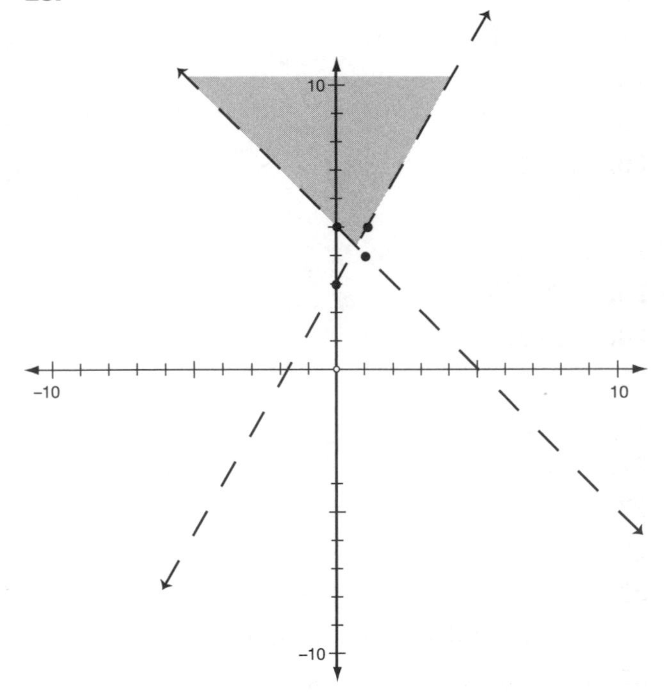

30.

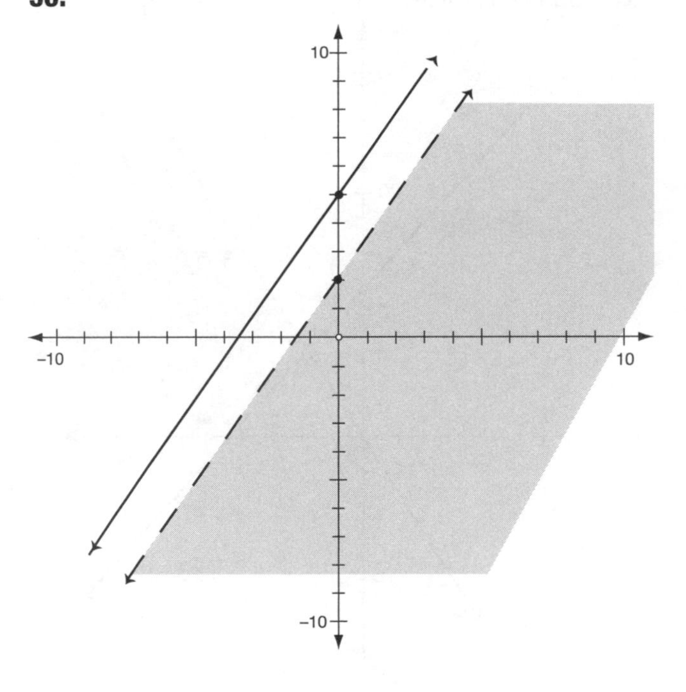

31. You need four $5 tickets and six $7 tickets.

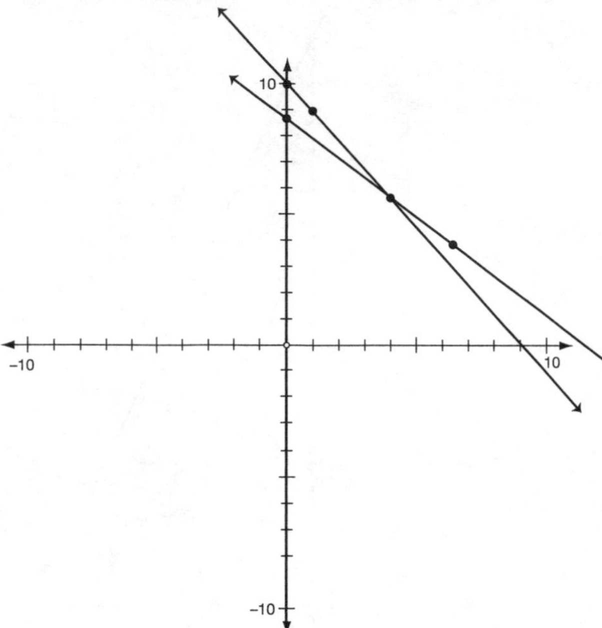

32. You need at least four girls in the group.

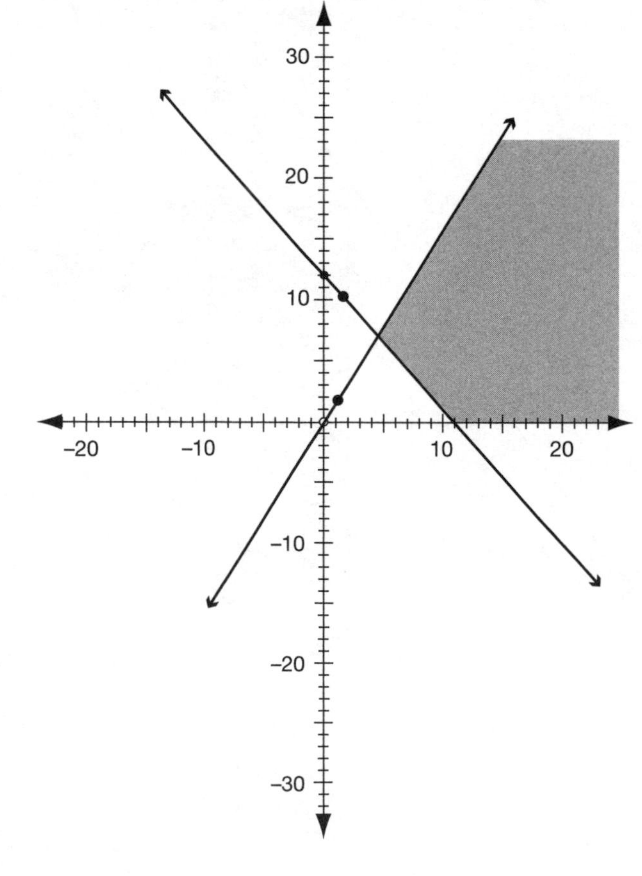

Lesson 12

1. (13,–5)
2. (10,3)
3. (2,–3)
4. ($\frac{1}{2}$,5)
5. (3,–1)
6. ($\frac{1}{2}$,–1)
7. (25,–12)
8. (1,–5)
9. (–8,17)
10. (4,5)
11. (3,2)
12. (5,–2)
13. (–2,–3)
14. (–1,0)
15. (2,10)
16. (0,2)
17. (–27,–11)
18. (–7,4)
19. (6,2)
20. (10,–2)
21. (2,8)
22. (1,3)
23. (5,2)
24. (–2,–4)
25. (2,4)
26. (2,–12)
27. (2,5)
28. (8,2)
29. (–1,1)
30. (0,–2)
31. 60 student tickets and 40 adult tickets
32. Four hours on the highways and two hours on the back roads

Lesson 13

1. $28x$
2. $3x + 4x^2$
3. a^{10}
4. $6x^2y$
5. $12x^3$
6. a^3b^5
7. $12m^6n^{11}$
8. $6r^3s^7$
9. $15a^4b^4c^2$
10. $23x^2y$
11. y^5
12. $\frac{1}{a^3}$
13. $-b^4$
14. a^2b
15. $5y$
16. $\frac{1}{9a^2b^4}$
17. $\frac{4xy^3}{z^2}$
18. $\frac{5s^2}{r}$
19. $\frac{4x^2}{5y^3}$
20. $\frac{m^3}{3}$
21. x^{10}
22. c^{20}
23. a^4b^6
24. x^3y^{15}
25. $m^{18}n^6$
26. $8x^9$
27. $9a^8b^4c^2$
28. $27x^3y^{15}z^{33}$
29. $16c^{10}d^{12}$
30. $16a^{12}b^4c^{20}$
31. $13x^5$
32. $42x^{10}$
33. $6x^3y^2$
34. $18x^2y^3$
35. $2x^9y^9$
36. $14x^2y^5$
37. $27a^{12}b^{15}$
38. 0
39. $6x^2y + 5xy^2$
40. $\frac{3a}{4b^2}$

Lesson 14

1. $5x - 5y + 10$
2. $7x^2 - 21x$
3. $24x^5 + 16x^4 - 40x^3$
4. $-6x + 6y + 42$
5. $3bx^2 + 5bxy + 3by$
6. $-14ac^2 + 35ac^3$
7. $6x^2y^2 - 6xy$
8. $6a^2x^2 - 4a^3bx + 20a^3x^4 + 16a^2x$
9. $-16x^4y^2 + 24x^3y^3$
10. $-8r^2st + 28r^3s^2 - 36rs^3t^2$
11. $x^2 + 6x + 8$
12. $x^2 + 3x - 18$
13. $x^2 - 8x + 12$
14. $x^2 + 9x - 10$
15. $x^2 + 2x + 1$
16. $2x^2 + 11x + 5$
17. $9x^2 - 1$
18. $20x^2 + 6x - 2$
19. $a^2 + 2ab + b^2$
20. $a^2 - b^2$
21. $12x^2 + 8x - 9xy - 6y$
22. $2x^2 - xy - y^2$
23. $30x^2 - 17x + 2$
24. $ac + ad + bc + bd$
25. $49y^2 - 4$
26. $16x^2 - 8x + 1$
27. $x^3 - 4x^2 + 5x -2$
28. $2x^3 + 7x^2 + 16x + 15$
29. $x^3 + 3x^2 - 4x - 12$
30. $3x^3 + 8x^2 - 5x - 6$

Lesson 15

1. $3(x + 4)$
2. $11(a + 3b)$
3. $3(a + 2b + 4c)$
4. $x(2x + 3)$
5. prime—can't be factored
6. $4x(4x + 5)$
7. $x(xy + 3)$
8. $2x(4x^2 - x + 2)$
9. $-6x^2(x - 3y)$
10. $10(x^4y^2 - 5x^3y + 7)$
11. $3a(2a - 13b)$
12. $4a^3bc(3b + a^2c)$
13. $11x^2y(2x^2 + 5y)$
14. $4(2x^2 + 3x + 5)$
15. $5(f^3 - 3f + 5)$
16. $5a^2b(6ab + 4 + 7a^2b)$
17. $(a + 7)(a - 7)$
18. $(b + 11)(b - 11)$
19. $(2x - 3)(2x + 3)$
20. prime—can't be factored
21. $(r + s)(r - s)$
22. $(6b + 10)(6b - 10)$
23. $(a^3 + b^3)(a^3 - b^3)$
24. $(y + 8)(y - 8)$
25. $(2x + 1)(2x - 1)$
26. $(5x + 2y)(5x - 2y)$
27. prime—can't be factored (because x^{25} is not a perfect square)
28. $(x^2 + 4)(x^2 - 4)$
29. $(b^5 - 6)(b^5 + 6)$
30. $(4a - 5b)(4a + 5b)$
31. $(x + 2)(x + 2)$
32. $(x + 4)(x + 2)$
33. $(x - 1)(x - 3)$
34. $(x + 4)(x + 3)$
35. $(x - 8)(x - 2)$
36. $(x - 14)(x - 1)$
37. $(x + 5)(x + 4)$
38. $(x - 10)(x - 2)$
39. $(x - 4)(x - 5)$
40. $(x - 6)(x - 5)$
41. $(x + 7)(x - 3)$
42. $(x + 3)(x - 2)$
43. $(x + 4)(x - 3)$
44. $(x + 5)(x - 2)$
45. $(x - 5)(x + 2)$
46. $(x - 8)(x + 1)$
47. $(x + 8)(x - 2)$
48. $(x - 7)(x + 3)$
49. $(x + 6)(x - 5)$
50. $(x - 6)(x + 3)$

Lesson 16

1. $(2x + 1)(x + 3)$
2. $(5x - 2)(x + 3)$
3. $(7x - 4)(x + 1)$
4. $(4x - 1)(x - 5)$
5. $(6x + 7)(x - 1)$
6. $(8x + 5)(x - 1)$
7. $(4x - 5)(2x + 1)$
8. $(3x - 1)(3x + 2)$
9. $(3x - 1)(2x - 5)$
10. $(5x + 3)(x - 2)$
11. $2(x^2 + 2)$
12. $(x - 2)(x + 2)$
13. $(x + 11)(x - 3)$
14. $(3x + 5)(3x - 5)$
15. $(x + 2)(x + 2)$
16. $2x(5x^3 + 6x - 3)$
17. $(11x - 1)(11x + 1)$
18. prime—can't be factored
19. $(x + 1)(x + 9)$
20. $(x + 5)(x - 3)$
21. $9(2x^2 + 39)$
22. $6a(a - 5b)$

23. $(7x-2)(7x+2)$
24. $(c-6)(c-5)$
25. $(a+b)(a-b)$
26. $(b+6)(b-3)$
27. $(n-7)(n+5)$
28. $6(4x+1)$
29. $(3x-10)(3x+10)$
30. $(5x-3)(x+2)$
31. $(3x+1)(2x-3)$
32. $(r-8)(r+3)$
33. $(f+9)(f-4)$
34. $3xy(x^2+2xy-3y^2)$
35. $(5x+1)(3x-2)$
36. $(5a+8)(5a-8)$
37. $6x^3y(8y^2-3x)$
38. $(2x+1)(3x+11)$
39. $5mn(2+mn^2-4m^2n)$
40. prime—can't be factored
41. $3(x+3)(x-3)$
42. $4(x+4)(x-4)$
43. $2(x+3)(x+3)$
44. $2(x-1)(x+3)$
45. $3(x+5)(x+2)$
46. $4(x^3+5)(x^3-5)$
47. $3(3x+5y)(3x-5y)$
48. $3(2x-7)(2x+1)$
49. $3(x-2)(x-6)$
50. $3(x+11)(x-2)$

Lesson 17

1. 6,–6
2. 5,–5
3. 7,–7
4. 5,–5
5. 7,–7
6. –2,1
7. –9,2
8. –9,5
9. –3,3
10. –8,–3
11. 9,–1
12. 1,1
13. 9,–5
14. $\frac{7}{2}=3\frac{1}{2},-1$
15. $-\frac{1}{3},7$
16. –2,1
17. –5,3
18. 2,–5
19. $-\frac{3}{4},\frac{3}{4}$
20. $\frac{2}{5},-6$
21. 3 feet
22. 5 ft. for the added length, $2\frac{1}{2}$ ft. for the walk
23. 2 in. for the added length, 1 in. for the width of the border
24. $2\frac{1}{2}$ ft.
25. 20 ft.

Lesson 18

1. 7
2. 9
3. 12
4. –8
5. 8
6. –6
7. a
8. 30
9. 40
10. 0
11. 0.2
12. $10x^2$
13. $-4a^4$
14. $15xy^9$
15. $-80a^3b$
16. $60x^2y$
17. $2\sqrt{2}$
18. $2\sqrt{5}$

19. $3\sqrt{6}$
20. $2\sqrt{10}$
21. $6\sqrt{2}$
22. $3\sqrt{3}$
23. $2\sqrt{7}$
24. $4\sqrt{10}$
25. $10\sqrt{2}$
26. $2\sqrt{6}$
27. 15
28. $10\sqrt{5}$
29. $20\sqrt{3}$
30. prime
31. $xy\sqrt{3}$
32. $2b^3$
33. $2c^2\sqrt{2d}$
34. $4abc^2\sqrt{5b}$
35. $2a^2b^3\sqrt{5ac}$
36. $10d^6\sqrt{5d}$
37. $\frac{1}{5}\sqrt{10}$
38. $\frac{x}{3}\sqrt{6}$
39. $\frac{ab}{2}\sqrt{2}$
40. $\frac{1}{7x}\sqrt{14x}$
41. $\frac{x}{3}\sqrt{15x}$
42. $\frac{2}{11}\sqrt{55}$
43. $\frac{3}{7}\sqrt{7}$
44. $\frac{6}{5}\sqrt{5}$
45. $\frac{1}{3}\sqrt{6}$
46. $\sqrt{2}$
47. $\frac{5}{x}\sqrt{x}$
48. $\frac{3}{2y}\sqrt{2y}$
49. $\sqrt{14}$
50. $2\sqrt{2a}$
51. $11\sqrt{7}$
52. $3\sqrt{3}$
53. $8\sqrt{2}$
54. $2\sqrt{2} - 2\sqrt{6}$
55. $17\sqrt{a}$
56. $5\sqrt{3} + \sqrt{5}$
57. $11\sqrt{x} - 4\sqrt{y}$
58. $5\sqrt{3}$
59. $12\sqrt{2}$
60. $5\sqrt{5} - \sqrt{7}$
61. $35\sqrt{6}$
62. $2\sqrt{3}$
63. $-12\sqrt{10}$
64. $2\sqrt{5}$
65. $12ab$
66. $2\sqrt{x}$
67. 6
68. 150
69. $60\sqrt{2}$
70. $\frac{15}{4}\sqrt{6}$

Lesson 19

1. ± 9
2. $\pm 5\sqrt{2}$
3. 64
4. 10.24
5. 4
6. 121
7. 25
8. 4
9. 4
10. 25
11. 46
12. 86
13. 47
14. 6
15. 25
16. 16
17. $-\frac{1}{3}$
18. $\frac{1}{81}$
19. 12
20. 4

Lesson 20

1. 4,8,1
2. 1,–4,10
3. 2,3,0
4. 6,0,–8
5. 4,0,–7
6. 3,0,0
7. 2,3,–4
8. 9,–7,2
9. 0,–2
10. 0,4
11. 5,–5
12. –5,1
13. –7,3
14. –5,–6
15. $-\frac{3}{2}$,–1
16. $\frac{1}{2}$, $\frac{1}{3}$
17. $\frac{-3 \pm \sqrt{5}}{2}$
18. $\frac{1 \pm \sqrt{11}}{2}$
19. $\frac{5 \pm \sqrt{17}}{2}$
20. $\frac{7 \pm \sqrt{61}}{2}$
21. $\frac{3 \pm \sqrt{29}}{2}$
22. $\frac{5 \pm \sqrt{41}}{8}$
23. $\frac{-11 \pm 5\sqrt{5}}{2}$
24. $\frac{-5 \pm \sqrt{37}}{2}$
25. 40 ft.

Posttest

If you miss any of the answers, you can find help for that kind of question in the lesson shown to the right of the answer.

1. b (1)
2. d (1)
3. c (1)
4. a (1)
5. d (2)
6. c (2)
7. c (2)
8. d (3)
9. c (3)
10. a (4)
11. b (4)
12. a (4)
13. b (4)
14. c (5)
15. c (5)
16. a (5)
17. d (6)
18. c (6)
19. c (7)
20. d (7)
21. b (8)
22. c (8)
23. c (8)
24. a (9)
25. a (9)
26. c (10)
27. d (10)
28. c (11)
29. a (11)
30. b (11)
31. d (12)
32. c (12)
33. c (13)
34. d (13)

35. b (14)
36. d (14)
37. c (15)
38. c (15)
39. d (16)
40. a (16)
41. c (17)
42. a (17)
43. c (18)
44. c (18)
45. b (18)
46. c (18)
47. d (19)
48. b (19)
49. a (20)
50. d (20)

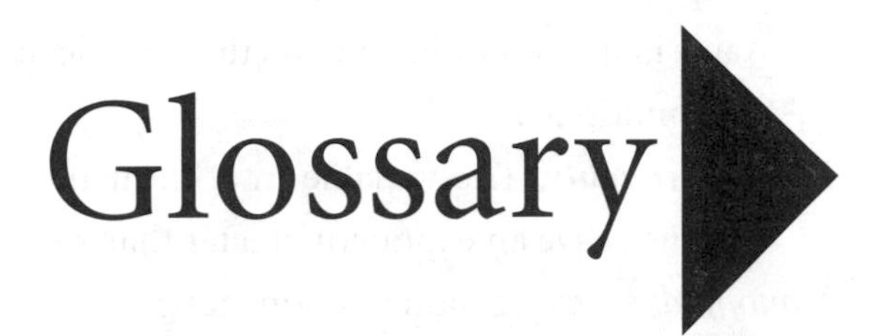

Glossary

Additive property of zero: When you add zero to a number, the result is that number.

Examples: $6 + 0 = 6$

$x + 0 = x$

Binomial: An expression with two terms.

Coefficient: The number in front of the variable(s).

Coincide: Occupy the same place in space.

Commutative property: Allows you to change the order of the numbers when you add or multiply.

Coordinate plane: Formed by two lines that intersect to form right angles.

Distributive property: Multiply the number and/or variable(s) outside the parentheses by every term inside the parentheses.

Examples: $2(a - b + 3) = 2a - 2b + 6$, $3x(x + 2) = 3x^2 + 6x$.

Empty set: A set with no members. When you solve an equation and there is no solution, your answer is the empty set. The symbol that represents the empty set is $\varnothing$.

Equation: Two equal expressions.

Examples: $2 + 2 = 1 + 3$, $2x = 4$.

Evaluate: Substitute a number for each variable and simplify.

Exponent: Tells you how many times a factor is multiplied. An exponent appears smaller and raised.

Example: $2^3 = 2 \cdot 2 \cdot 2$.

Factors: Numbers to be multiplied.

Example: Factors of 12 are 2 times 6.

Formula: A special equation that shows a relationship between quantities.

Identity: When the variables in an equation are eliminated and the result is a true statement, you will have an infinite number of solutions. Because every real number is a solution, we use the notation, R, to represent the set of real numbers.

Inequality: Two expressions that are not equal and are connected with an inequality symbol such as $<$, $>$, $\leq$, $\geq$, or $\neq$.

Infinite: Endless.

Integers: All the whole numbers and their opposites. Integers do not include fractions. The integers can be represented in this way: ... –3, –2, –1, 0, 1, 2, 3, ...

Like terms: Terms that have the same variable(s) with the same exponent.
Example: $3x^2y$ and $5x^2y$.

Linear equation: Always graphs into a straight line. The variable in a linear equation cannot contain an exponent greater than one. It cannot have a variable in the denominator, and the variables cannot be multiplied.

Linear inequality: The variable in a linear inequality cannot have an exponent greater than one.

Monomial: An expression with one term.

Multiplicative inverse: You get the multiplicative inverse by inverting the number. A number times the multiplicative inverse will equal 1.
Examples: $2 \cdot \frac{1}{2} = 1, \frac{3}{4} \cdot \frac{4}{3} = 1, -\frac{1}{5} \cdot -5 = 1,$ and $-\frac{2}{3} \cdot -\frac{3}{2} = 1$.

Order of operations: The order of performing operations to get the correct answer. The order you follow is:

1. Simplify all operations within grouping symbols such as parentheses, brackets, braces, and fraction bars.
2. Evaluate all exponents
3. Do all multiplication and division in order from left to right.
4. Do all addition and subtraction in order from left to right.

Ordered pair: A pair of numbers that has a specific order. The numbers are enclosed in parentheses with the x coordinate first and the y coordinate second.
Example: (2,3).

Origin: On a number line, the origin is your starting point or 0. On a coordinate plane, the origin is the point where the two intersecting lines intersect. The coordinates of the origin are (0,0).

Polynomial: A number, variable, or combination of a number and a variable.
Examples: 5, $3x$, and $2x + 2$.

Prime: A prime number is a number that cannot be factored further. The only factors of a prime number are one and the number itself. Examples: 2, 3, and 5.

Quadrants: The coordinate plane is divided into 4 equal parts called quadrants. A number names each quadrant. The quadrant in the upper right-hand quadrant is quadrant I. You proceed counterclockwise to name the other quadrants.

Quadratic equation: An equation where the highest power of the variable is 2. The graph of a quadratic equation is a smooth curve. A quadratic equation will always have two solutions. A quadratic equation is represented by: $ax^2 + bx + c = 0$.

Quadratic formula: $\frac{-b \pm \sqrt{b^2 - 4ac}}{2a}$

Radical equation: An equation that has a variable in the radicand.

Radical sign: The mathematical symbol that tells you to take the root of a number.
Example: $\sqrt{}$

Radicand: The number under the radical sign in a radical. In the radical $\sqrt{16}$, the radicand is 16.

Simplify: Writing a number or expression in its simplest form.
Example: $\frac{5}{10} = \frac{1}{2}$ and $2x + 3x = 5x$.

Slope: The steepness of a line. Slope is also the rise over the run or the change in y over the change in x. Slope can be calculated by using the formula: $\frac{y_2 - y_1}{x_2 - x_1}$.

Slope-intercept form: $y = mx + b$. Also known as *y = form.*

Square root: The opposite of squaring. The square root of 16 is 4 because 4 times 4 equals 16. The mathematical symbol that tells you to take the square root of 16 is $\sqrt{16}$.

Squaring a number: Multiplying a number by itself.
Example: $4 \cdot 4$.

System of equations: Two or more equations with the same variables.

System of inequalities: Two or more inequalities with the same variables.

Term: Terms are separated by addition and subtraction signs. The expression $a + b$ has two terms. The expression ab has one term.

Trinomial: An expression with three terms.
Example: $a + b + c$.

Variable: A letter representing a number.

Whole numbers: 0, 1, 2, 3, … Whole numbers start with 0 and do not include fractions.

x-axis: The horizontal line that passes through the origin on the coordinate plane.

y-axis: The vertical line that passes through the origin on the coordinate plane.

y-intercept: Point where the line intersects the y-axis.

Zero product property: When the product of two numbers is zero, then one or both of the factors must equal zero.
Example: $ab = 0$ if $a = 0$, $b = 0$, or both equal 0.

Additional Resources

There are many resources available to help you if you need additional help or practice with algebra. Your local high school is a valuable resource. Most high school math teachers would assist you if you asked them for help with a lesson. If you need a tutor, the teacher may be able to suggest one for you.

Colleges are also a valuable resource. They often have learning centers or tutor programs available. To find out what is available in your community, call your local college's math department or learning center.

If you would like to continue working algebra problems on your own, there are books available at your local bookstore or library that can help you. The following algebra textbooks and workbooks provide helpful explanations or practice sets of problems. Check your local bookstore, library, or high school to see if they are available.

Algebra the Easy Way, Fourth Edition, by Douglas Downing, 2003, Barron's Educational Series
Algebra I (Cliff's Quick Review), by Jerry Bobrow, 2001, Cliff's Notes
Algebra for the Clueless, by Robert Miller, 1998, McGraw-Hill

NOTES

NOTES

NOTES

Special offer from LearningExpress!

Let LearningExpress help you acquire practical, essential algebra skills FAST!

Go to LearningExpress Practice Center at www.LearningExpressFreeOffer.com, an interactive online resource exclusively for LearningExpress customers.

Now that you've purchased LearningExpress's *Algebra Success in 20 Minutes a Day* skill-builder book, you have **FREE** access to:

- **Four exercises covering ALL VITAL ALGEBRA SKILLS**, from dealing with word problems to figuring out odds and percentages
- **Immediate scoring** and **detailed answer explanations**
- Benchmark your skills and focus your study with our **customized diagnostic report**
- **Improve** your math knowledge and **overcome math anxiety**

Follow the simple instructions on the scratch card in your copy of *Algebra Success.* Use your individualzed access code found on the scratch card and go to www.LearningExpressFreeOffer.com to sign-in. Start practicing your math skills online right away!

Once you've logged on, use the spaces below to write your access code and newly created password for easy reference:

Access Code: ____________________ Password: ____________________

NOTES